Cities at Risk

Advances in Natural and Technological Hazards Research

Volume 33

For further volumes:
http://www.springer.com/series/6362

Helene Joffe • Tiziana Rossetto • John Adams
Editors

Cities at Risk

Living with Perils in the 21st Century

Springer

Editors
Helene Joffe
Division of Psychology and Language Sciences
University College London
London, United Kingdom

Tiziana Rossetto
Department of Civil, Environmental and Geomatic Engineering
University College London
London, United Kingdom

John Adams
Department of Geography
University College London
London, United Kingdom

ISSN 1878-9897 ISSN 2213-6959 (electronic)
ISBN 978-94-007-6183-4 ISBN 978-94-007-6184-1 (eBook)
DOI 10.1007/978-94-007-6184-1
Springer Dordrecht Heidelberg New York London

Library of Congress Control Number: 2013933344

Printed on acid-free paper

Springer is part of Springer Science+Business Media (www.springer.com)

Acknowledgements

We would like to thank Cliodhna O'Connor for her generous and insightful help in editing this book. We would also like to thank the production team at Springer and all of the contributors for producing what we hope will be a very informative and useful book.

Contents

Contributors

John Adams Department of Geography, University College London, London, UK

Julia Becker Joint Centre for Disaster Research, Massey University/GNS Science, Wellington, New Zealand

Tim Harries Behaviour and Practice Research Group, Business School, Kingston University London, Kingston-upon-Thames, UK

Miriam Hughes Joint Centre for Disaster Research, Massey University/GNS Science, Wellington, New Zealand

Huraera Jabeen Department of Architecture, BRAC University, Dhaka, Bangladesh

Helene Joffe Division of Psychology and Language Sciences, University College London, London, UK

Cassidy Johnson Development Planning Unit, Faculty of the Built Environment, University College London, London, UK

David Johnston Joint Centre for Disaster Research, Massey University/GNS Science, Wellington, New Zealand

Nuray Karanci Department of Psychology, Middle East Technical University, Ankara, Turkey

Graham Leonard Joint Centre for Disaster Research, Massey University/GNS Science, Wellington, New Zealand

Michael Lindell Hazard Reduction and Recovery Center, Texas A&M University, College Station, TX, USA

Sara McBride Joint Centre for Disaster Research, Massey University/GNS Science, Wellington, New Zealand

John McClure School of Psychology, Victoria University, Wellington, New Zealand

Cliodhna O'Connor Division of Psychology and Language Sciences, University College London, London, UK

Douglas Paton School of Psychology, University of Tasmania, Hobart, Launceston, TAS, Australia

Tiziana Rossetto EPICentre, Department of Civil, Environmental and Geomatic Engineering, University College London, London, UK

John Twigg Department of Civil, Environmental and Geomatic Engineering, University College London, London, UK

Kim Wright Joint Centre for Disaster Research, Massey University/GNS Science, Wellington, New Zealand

Katsuya Yamori Disaster Prevention Research Institute, Kyoto University, Gokasho, Uji, Kyoto, Japan

Chapter 1
Introduction – Living with Perils in the Twenty-First Century

Tiziana Rossetto

Abstract This chapter introduces some of the questions that spurred the writing of this book, particularly whether there are underlying commonalities in beliefs about natural hazards across the globe, and whether links can be drawn between people's perceptions or representations of their risk from natural hazards on the one hand, and the actions they take to mitigate against these hazards on the other. The book brings together a number of risk perception/representation studies that traverse several countries and hazards and are authored by academics from a diverse set of disciplines. Challenges and gaps are identified in relation to the risk perception field, and it is argued that cross-cultural, multi-disciplinary studies are required to answer the questions raised in the chapter.

1.1 Natural Hazards in the Twenty-First Century

Since the mid-to-late twentieth century, the world has seen an increase in the number and economic cost of disasters resulting from natural hazards (see Fig. 1.1). Many studies (e.g. Nellemann et al. 2008) have shown that there has been an increased rate of occurrence of climatic events (e.g. cyclones and floods) over the last three decades, a trend that may partly be explained by climate change. Figure 1.1, however, also shows that some of the worst disasters have been a result of earthquakes, for which there is no evidence of an increased occurrence rate. This implies that the consequences (in terms of affected population and cost) of some natural disasters are increasing due to human-made vulnerabilities.

T. Rossetto (✉)
EPICentre, Department of Civil, Environmental and Geomatic Engineering,
University College London, Gower Street, London WC1E 6BT, UK
e-mail: t.rossetto@ucl.ac.uk

H. Joffe et al. (eds.), *Cities at Risk: Living with Perils in the 21st Century*,
Advances in Natural and Technological Hazards Research 33,
DOI 10.1007/978-94-007-6184-1_1, © Springer Science+Business Media Dordrecht 2013

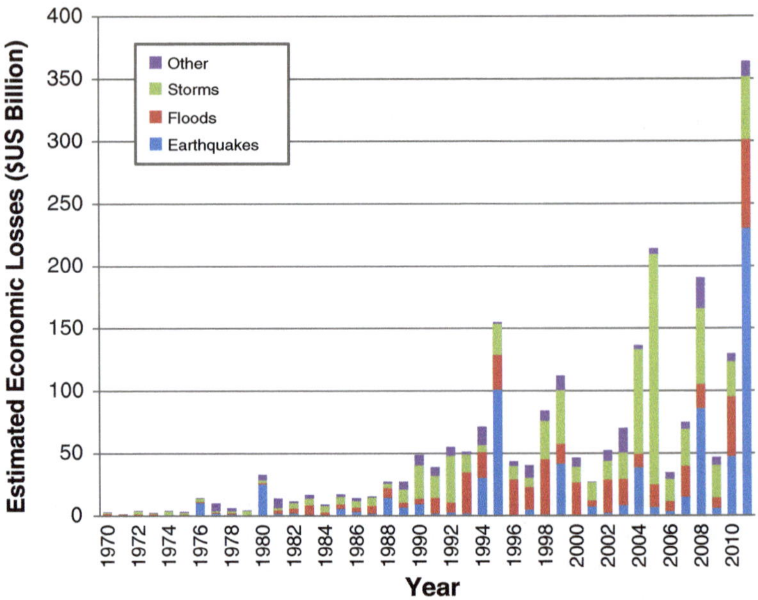

Fig. 1.1 Estimated economic losses ($US billion) due to natural disasters from 1975 to 2011. The category "Other" includes wildfires, droughts, extreme temperatures and volcanoes (*Source*: EM-DAT 2012)

The world's population has increased from about 1.6 to 6.1 billion people over the last century, and modern trends in living have led to rapid urbanisation, often accompanied by a rapid increase of poorly built housing, uncontrolled use of land, occupation of unsafe environments, overstretched services and high population densities (Alexander 2000; Blaikie et al. 2003). Simply put, when a natural hazard hits an urban area there are now more (vulnerable) people and urban assets exposed, leading to greater numbers of affected people and higher economic losses.

In recent years the economic losses associated with natural hazard occurrences have also escalated due to globalisation. For example, the 2011 floods in Thailand, which affected 65 out of 77 provinces and lasted around 4.5 months, became one of the five most costly disasters ever recorded (with a cost of $US 45.7 billion estimated by the World Bank [World Bank 2011]), with damage to Thailand's electrical and car manufacturing industries leading to global supply chain disruptions.

Despite the increasing occurrence of disasters and the increasing number of people affected by them (i.e. made homeless or injured), it is interesting to note that the number of deaths due to natural hazards is reportedly decreasing (EM-DAT 2012). This is especially true in developed countries (Khan 2005). In the disaster literature, this is attributed to improvements in science, engineering, medicine, communications, disaster preparedness, mitigation and management over the last century (e.g. Alexander 1999; Guha-Sapir et al. 2004; Lindell and Prater 2003). Taking my own field of earthquake engineering as an example, since the late nineteenth century we have evolved from not even knowing that earthquakes were associated with fault ruptures and plate-tectonics, to technologically advanced devices to reduce earthquake-induced movements in structures, and to designing structures to be damaged in "life-safe" ways.

Although we now have more knowledge on natural hazards, on how disasters happen and how they can be mitigated, the implementation of disaster risk reduction practices remains a challenge. For example, in the case of earthquake engineering, although seismic building codes exist in many countries, their application sometimes fails due to a number of factors such as lack of risk awareness, finances, enforcement, and technical capability (amongst others). Even in the USA and Japan, two highly developed countries where seismic risk awareness is high and governance and enforcement are effective, Rossetto et al. (2011) and Joffe et al. (in press) show that individual households do little to mitigate against a potential future earthquake event.

The existing psychological literature provides conflicting observations on what drives mitigative behaviour in the case of natural hazards (see Solberg et al. 2010). This book aims to shed light on this issue, in the hope that it may be possible to act on this knowledge to more effectively promote mitigation measures against the perils of the twenty-first century.

1.2 The Book

On the 26th October 2009, an international risk perception symposium, "Cities at Risk: Living with Perils in the 21st Century," was held at University College London (UCL). The conference was organised jointly by UCL EPICentre and the Joint Centre for Disaster Research (JCDR) of Massey University in New Zealand, and led to two subsequent symposia in New Zealand in 2010 and 2011. The aims of the symposia were to:

- Present and discuss state-of-the-art international research in the field of risk perception of natural hazards;
- Explore the value of multi-disciplinary research in this field;
- Explore the challenges involved in the cross-cultural study of risk perception;
- Suggest alleys for future research in risk perception.

This book gathers contributions from a range of authors who have participated in these conferences. It collects case studies from around the world, examining how people respond to the natural hazards that they face. By "respond to" we mean both how they perceive/represent these hazards and how they behave in anticipation of them. In particular, work that looks at the links between perceptions/representations and behaviours has been specifically included.

Authors are intentionally drawn from different disciplinary backgrounds, ranging from Engineering to Psychology, Geography, Anthropology and Urban Planning. The book therefore not only brings to light the ways that different cultures represent and perceive natural hazards but also the different ways in which various disciplines write about living with perils in the twenty-first century. It is hoped that each discipline can learn and draw insights from the other perspectives displayed in the book.

The book begins with a chapter by Joffe and O'Connor that provides an overview of developments in the risk perception field, followed by a chapter by Adams that presents some framing devices that can be used for the study of risk across cultures. These chapters offer alternative perspectives on risk to those usually seen in the risk perception field, thereby forging pathways through which more valid models of how publics engage with risks can be built.

A series of chapters presenting research that involves different countries and addresses responses to both single and multiple hazards then follow. These chapters cover Bangladesh, Japan, New Zealand, Turkey, the USA and the UK, and include a range of natural hazards: floods, earthquakes, tornadoes, hurricanes, volcanoes, with the Bangladeshi case also presenting salient effects of realised climate change. All contributors were asked to address the following specific dynamics in some way: fatalism, trust, worry/anxiety, responsibility, and sense of vulnerability (in a social scientific sense). The chapters thereby allow for observation of how these concepts manifest in relation to different hazards, cultural contexts, and academic disciplines.

It is interesting to note that as this book was being compiled, further large events occurred in two geographical areas included in the book, namely the 2010 Christchurch earthquake in New Zealand and the 2011 Tōhoku earthquake and tsunami in Japan. The chapters that focus on New Zealand and Japan (Johnston et al. and Yamori, respectively) offer interesting comparisons of local orientations to risk before and after these events.

The contributing authors approach risk from different theoretical positions, particularly risk perception (e.g. Jabeen & Johnson, Lindell) and risk representation (e.g. Harries, Joffe & O'Connor, and Yamori). Some authors are more focused on risk related behaviours (e.g. Lindell), whilst others also introduce perspectives on resilience (e.g. Adams, Karanci, Jabeen & Johnson). A range of methods are used to gather lay-people's responses and derive meaning from them. Despite this diversity of cultures and methods, some similarities can be identified in the ways that people respond to the hazards they face. These commonalities are drawn together and highlighted in the final chapter by Twigg, who also relates them to current practices in the communication of risk and disaster risk management.

Twigg's contribution makes clear that risk perception/representation studies have an important role to play in the shaping of disaster risk reduction initiatives. In actual practice, however, there seems to be little evidence that such studies are indeed used for this purpose. This is partly due to the way in which risk perception/representation studies have been carried out to date and what they have measured. Twigg provides a detailed discussion of this. However, it is thought useful to present here some of the ways that risk studies of the future could be designed to optimise their impact on disaster risk reduction, such that the reader may have these in mind whilst reading the book.

1.3 Gaps and Challenges in Risk Perception Studies in the Twenty-First Century

The chapters in this book clearly present distinct views on how natural hazard risks are seen in different cultures. However, careful reading shows that there are also many common strands to people's perceptions/representations and an evolution in the methods used to study this subject. Twigg suggests that a multi-disciplinary approach to the study of risk perception and representation in the twenty-first century is essential not only for furthering knowledge but also for developing more effective disaster risk reduction initiatives.

Systematic cross-cultural studies on risk perception and risk representation are rare, with two exceptions in the case of earthquake hazard being Palm and Carroll (1998) and Joffe et al. (in press). Cross-disciplinary studies also seldom materialise in the literature. Such studies present many challenges (some of which are summarised in Table 1.1). However, none of these are thought to be insurmountable, and it is to be hoped that the twenty-first century will see many such studies enriching the literature.

Twigg highlights several other salient gaps in the current risk perception field, notably longitudinal studies, studies conducted in developing countries (i.e. the "global South"), and studies of stakeholders in risk reduction other than lay-people (e.g. disaster risk managers, members of local authorities).

In terms of longitudinal studies, Lindell shows that several studies of risk perception regarding natural hazards have been carried out at different times in the USA. However, he states that these cannot be built into an effective longitudinal comparison as different groups/types of respondents have been used as well as different metrics and tools for measuring risk perception, making the studies not directly comparable. In this book Yamori and Johnston et al. present rare examples of comparisons in lay perspectives on risk before and after major natural disasters. Ideally, more such studies that explore the evolution of risk representation/perception through the entire disaster cycle–i.e. pre-, mid- and post-event through to the next event - would be highly desirable. Such longitudinal studies would present opportunities for cross-disciplinary collaboration to understand how the experience of the event, recovery and reconstruction, risk communication, governance, and mitigation promotion initiatives influence people's perceptions and preparatory behaviours. Thus, they would provide a significant contribution to future disaster risk reduction initiatives. Although such

Table 1.1 Non-exhaustive list of challenges (presented as questions) in designing and achieving cross-cultural multi-disciplinary studies of risk

Cross-cultural studies	Multi-disciplinary studies
What data collection methods are relevant to the study of multiple cultures?	How do we communicate effectively across disciplines?
What factors/parameters/indicators should/can be used to compare cultures?	What is the most useful language/terminology?
How can responses be related to contexts?	How do we deal with differing study objectives?

studies may seem difficult to imagine for infrequent hazards (like earthquakes), they would be more feasible for frequently occurring hazards (e.g. flooding).

Risk perception/representation studies for developing countries are rare, and again no longitudinal studies exist for these areas. Why is this case? Perhaps some studies exist but are not published in the international literature, perhaps the relevance of such studies is not appreciated in some countries, or perhaps there are few people trained in this field. Nevertheless, it is in these locations that the study of risk perception/representation and their interplay with preparatory behaviours have most to contribute. As Jabeen and Johnson show, it is in these societies that the individual has a key role in their own protection against natural hazards, and where improvements in the communication of risk and uptake of mitigation measures can have a tangible impact on saving lives.

The majority of existing risk perception studies are carried out on individuals in the community, rather than community groups. Yet, as highlighted in Johnston and colleagues' and Karanci's chapters, community groups play an important role in disaster risk reduction activities and what drives the actions of a group can be very different from those drivers pertaining to individuals. Equally, few studies look at stakeholders other than lay-people and home-owners. A study of the risk representation and decision environment of key policy makers involved in disaster risk reduction would be beneficial for understanding how decisions in this field are made. Of particular interest would be a study of a number of policy/decision makers with different temporal scales of power/activity, at times preceding, during and following a natural disaster e.g. members of local authorities in office for short or long periods throughout the disaster cycle, or disaster managers who take on active roles only in the case of disasters.

Although the book includes chapters on long and short-return period natural hazard events, there has been no systematic analysis of how risk perceptions vary between these. Jabeen and Johnson state that mitigation measures adopted against frequent hazard events in Bangladesh may also be of use to protect against the effects of rarer events. If this is also true of other parts of the world, then such studies could inform the future design of mitigation campaigns, which might target multiple hazards and improve the currently poor uptake of mitigation measures for infrequent hazards.

1.4 Final Remarks

In light of predicted changes in the frequency and severity of natural hazards over the twenty-first century, it has never been more important to understand how risk perception/representation studies can contribute to the overall reduction of hazard effects. The cases presented in the following chapters provide significant insight into the current state-of-art in risk perception research across several disciplines. It is hoped that the contributions contained in this book will provide inspiration for future directions in risk perception research.

References

Alexander D (1999) Natural disasters. Kluwer Academic Publishers, Dordrecht

Alexander D (2000) Confronting catastrophe: new perspectives on natural disasters. Oxford University Press, New York

Blaikie P, Cannon T, Davis I, Wisner B (2003) At risk: natural hazards, people's vulnerability, and disasters, 2nd edn. Routledge, London

EM-DAT (2012) EM-DAT: the OFDA/CRED international disaster database. www.emdat.be. Accessed 25 Oct 2012

Guha-Sapir D, Hargitt D, Hoyois P (2004) Thirty years of natural disasters 1974–2003: the numbers. Presses Universitaires de Louvain, Louvain-la-Neuve

Joffe H, Solberg C, Rossetto T, O'Connor C (2013) Social representations of earthquakes: a study of those living in three highly seismic areas. Earthquake Spectra

Khan ME (2005) The death toll from natural disasters: the role of income, geography and institutions. Rev Econ Stat 87(2):271–284

Lindell MK, Prater CS (2003) Assessing community impacts of natural disasters. Nat Hazards Rev 4(4):176–185

Nellemann C, Hain S, Alder J (eds) (2008) In dead water: merging of climate change with pollution, over-harvest, and infestations in the world's fishing grounds. United Nations Environment Programme, GRID-Arendal, Arendal

Palm R, Carroll J (1998) Illusions of safety. Culture and earthquake hazard response in California and Japan. Westview Press, Boulder

Rossetto T, Solberg C, Joffe H (2011) A different view on human vulnerability to earthquakes: lessons from risk perception studies. In: Spence RJ, So E, Scawthorn C (eds) Human casualties in earthquakes: progress in modeling and mitigation, vol 29. Springer, Dordrecht, pp 291–304

Solberg C, Rossetto T, Joffe H (2010) The social psychology of seismic hazard adjustment: re-evaluating the international literature. Nat Hazards Earth Syst Sci 10(8):1663–1677

World Bank (2011) The world bank supports Thailand's post-floods recovery effort. http://www.worldbank.org/en/news/2011/12/13/world-bank-supports-thailands-post-floods-recovery-effort. Accessed 18 Oct 2012

Chapter 2
Risk Society and Representations of Risks: Earthquakes and Beyond

Helene Joffe and Cliodhna O'Connor

Abstract This chapter explores how lay publics respond to potential disasters. It contends that the current risk perception field largely neglects the common-sense beliefs and emotions that lie at the root of public responses to risks. The chapter challenges several of the assumptions that buttress the conventional construal of the terms 'risk' and 'perception'. It proposes that the current focus on how the individual mind cognitively processes predictable, calculable phenomena should be replaced by emphasis on how emotional and socio-cultural beings represent often unknowable potential catastrophes. Social representations theory is put forward as a viable theoretical framework within which this shift could be achieved. The chapter illustrates the value of a social representations approach to studying risk by presenting the findings of a cross-cultural study examining social representations of earthquakes in cities at risk of earthquakes in the US, Japan and Turkey. The chapter concludes by proposing routes by which the findings of such studies could be channelled into behavioural intervention programmes.

2.1 Introduction

This chapter examines how members of the public represent potential catastrophes, from earthquakes to pandemics. It offers a challenge to the 'risk perception' research agenda as well as to the conventional construal of both of the terms 'risk' and 'perception', en route to proposing core emotional and socio-cultural drivers that shape how people face the dangers and catastrophes of the future. In doing so, it complements and extends existing work on how people think, feel and act in relation to the host of dangers that they face.

H. Joffe (✉) • C. O'Connor
Division of Psychology and Language Sciences, University College London, Gower Street, London WC1E 6BT, UK
e-mail: h.joffe@ucl.ac.uk; cliodhna.oconnor.10@ucl.ac.uk

H. Joffe et al. (eds.), *Cities at Risk: Living with Perils in the 21st Century*,
Advances in Natural and Technological Hazards Research 33,
DOI 10.1007/978-94-007-6184-1_2, © Springer Science+Business Media Dordrecht 2013

2.2 Challenging the Term 'Risk'

The contemporary definition of risk used by experts relates to predicting the "probability of an event, combined with the magnitude of losses and gains that it will entail" (Douglas 1992, p. 40). In the safety sphere the 'gains' element tends to drop away and 'risk' represents the risk of death, but also the risk of disfigurement or disability. Thus, regarding the perils of the future, risk can simply be defined as the probability that a harmful event will occur (Weinstein 2003).

The emphasis on probability calculation demonstrates that this understanding of risk represents a very different orientation to peril from that which existed in pre-modern times (Bernstein 1998). In the anthropologist Mary Douglas' accounts of pre-modern societies, risk simply means 'danger from future damage', with a community's construal of the dangers they face intimately linked to local cultural operations of morality, responsibility attribution and intergroup boundaries (Douglas 1992; Douglas and Wildavsky 1982). In contrast, contemporary usage of the term 'risk' implies precision of calculation, objectivity and control (Joffe 1999). Viewing dangers in terms of causal, predictable relations is a distinctively modern phenomenon, swathing dangers in an aura of scientific measurement and control (Bernstein 1998; Douglas 1992).

Undoubtedly, modern scientific and technological advances have indeed increased the extent to which human society can control many dangers. A host of diseases that blighted previous eras, for example, have been rendered harmless due to use of antibiotics and vaccination programmes. However, a key dimension of risk in contemporary society is that the very interventions implemented to manage risks can set off unknowable chain reactions that generate further perils that elude human control. The emergence of new antibiotic-resistant strains of bacteria due to antibiotic overuse is a case in point. This observation buttresses the contention of sociologist Ulrich Beck (1992, 2006) that the sense of control over perils that is implied by the contemporary construal of 'risk' is largely illusory and ironic. Beck terms our late modern society 'risk society' (Beck 1992), or more latterly 'world risk society' (Beck 2006), a phrase that "describes a phase of development of modern society in which the social, political, ecological and individual risks created by the momentum of innovation increasingly elude the control and protective institutions of industrial society" (Beck 1996, p. 27).

Beck postulates that contemporary society is characterised by a profusion of fundamentally incalculable risks and that within the risk society, highly developed institutions (most notably science) try to anticipate what cannot be anticipated. Beck illustrates this irony with the following example: "when suicide terrorists succeeded in turning commercial passenger aircraft into rockets, which destroyed symbols of American world power" this was "an action that was utterly improbable according to every logic of risk" (Beck 2006, p. 330). The irony here, in terms of the science of predicting risk, is that past experience encouraged anticipation of the wrong type of risk: before September 11, 2001, the World Trade Centre had been the target of a terrorist truck bomb, making camera surveillance widespread but disregarding the possibility of a potential air attack. This instance of relying on historical risk patterns to foretell those of the future stands as a sobering example for those working within the field.

A more recent manifestation of the unpredictable nature of danger lay in the unanticipated outbreak of widespread Swine Influenza (H1N1) between the summer of 2009 and autumn of 2010, which arose in place of a predicted Highly Pathogenic Avian Influenza (HPAI) pandemic. One might argue that the World Health Organisation's (WHO) declaration of (what turned out to be) a relatively mild bout of Swine Influenza as a pandemic owed more to its historical links with the 1918 'Spanish Flu' (which may have contained strains of the same H1N1 virus) and the anticipation of severe strains of Highly Pathogenic Avian Influenza than to the outbreak itself. Extensive preparations had been made for HPAI, including efforts by the WHO and UNICEF to devise programmes for drug and vaccine supply as well as public education and engagement activities (Scoones and Foster 2010). These plans were quickly adopted as the basis for managing the unexpected H1N1 outbreak, with the result that the major international effort to produce and distribute anti-retrovirals and vaccinations was directed at a more severe level of risk than that which arose. The science of predicting risk can therefore prove an unreliable guide for risk management decisions, causing scientific and financial resources to be channelled into activities that are wide of the mark.

A final example of past experience encouraging anticipation of the wrong type of risk relates to the devastation that followed the March 2011 earthquake and tsunami in Tōhoku, Japan. For economic reasons, most countries choose not to produce building codes and structures that anticipate the most extreme seismic scenarios (and consequently are most costly). Rather, authorities determine the levels of structural resilience to which buildings must conform based on what are considered to be the *most likely* upper limits of future earthquakes and tsunami. These calculations are made on the basis of the seismic history of the area. On the basis of probabilistic assessments of the hazard in each location, which take into account earthquake magnitude over previous centuries, Japanese engineers had prepared structures (in terms of building codes and coastal defences) to withstand earthquakes of a magnitude between 7 and 8 (Nishikawa 2011) and their equivalent tsunami. Disastrously, the earthquake that hit Tōhoku on March 11, 2011 measured magnitude 9 and, together with its incipient tsunami, destroyed villages, damaged several nuclear reactors and caused the deaths of approximately 16,000 people, with a further 3,000 recorded as missing (World Health Organisation 2012). Preparing solely for the statistically probable can therefore produce catastrophic results.

These three examples indicate that far from being foreseeable and calculable phenomena, many future dangers are unknown and unknowable. Complex, long return period,[1] spatially variable dangers – ranging from earthquakes to pandemics to climate change – often cannot be accurately predicted. Those prediction efforts that are

---

[1] The term 'return period' refers to the average intervals of time between hazard events such as floods or earthquakes.

nevertheless mounted may not produce very useful information: subject to numerous sources of error and variation, they are at best merely "pieces of an intricate puzzle that may sometimes contribute to improved decisions" (Sarewitz 2010, p. 24). The gradual devaluation of prediction is vividly illustrated in the historical evolution of the field of seismology, which has (rather uniquely among scientific fields) decisively shifted its focus away from prediction towards building resilience in the face of vulnerability (Hough 2010; Sarewitz 2010). Indeed, earthquake prediction is now a stigmatised pursuit among seismologists (Joffe et al. submitted; Joffe 2013). Sarewitz (2010) argues that the climate change field should similarly aspire to tackling people's vulnerability rather than predicting future climate scenarios. Thus, the validity and even desirability of treating risks as predictable, calculable phenomena is challenged by researchers, practitioners and theorists in certain fields.

2.3 Challenging the Term Risk 'Perception'

Beck's risk society theory effectively critiques the often inflated confidence of scientific fields in their powers of calculability. Once 'risks' are reframed as partly unknowable dangers or perils, rather than predictable, calculable entities, this chapter argues that we need a better model of what members of the public do with the unknowable than we currently see in the 'risk perception' field. The term 'perception' generally pertains to sensorial knowledge – how one's senses register a particular entity. Since the 1950s, the study of risk perception has been dominated by the field of cognitive psychology, which generally approaches risk perception as a form of 'information processing', focusing on the internal cognitive processes that occur when laypeople are confronted with risk.

In recent times, the validity of conceptualising risk perception solely in terms of individual cognition has been questioned. One particularly salient basis for critique argues that treating responses to risk as mechanical information-processing that occurs within the individual mind obscures the inter-subjective aspects of knowledge (Joffe 2003). People do not navigate the world as atomised individuals, but rather are embedded in social groups, cultures and societies. When confronted with risks, people draw on the meaning systems and ways of thinking associated with the groups to which they belong; responses to risks develop in and through interaction with other social actors (Kahan et al. 2010; Joffe 2003). In contemporary society, risks are generally 'perceived' not via direct sensory information, but via media reports heralding news of an imminent danger. Such reports often relay scientifically-generated information in a more populist form, for example, sensationalised media coverage of emerging infectious diseases (Joffe and Haarhoff 2002; Washer 2004, 2006; Washer and Joffe 2006; Washer et al. 2008). The formation of 'risk perception' does not involve independent observation of stable external objects, but rather handling an intricate mix of visual and textual information that is communicated by a host of other social agents and institutions. Thus, what is termed 'perception' is a complex process involving engagement with sets of images and words that circulate in society and that may re-interpret 'official' risk information. Focusing one's lens purely on the individual mind means that the constitutive influence of such social phenomena remains unexplored.

A further area that remains largely untapped by conventional understandings of risk perception relates to the emotional dimension of responses to risk (for exceptions see Lerner et al. 2003; Peters et al. 2004; Sjöberg 2007). Risk perception is often characterised by rather 'cold' cognitive variables, such as estimated likelihood of being affected by the threat, perceived efficacy in relation to it and perceived severity of the risk. While research on such variables has undoubtedly furnished useful results, an exclusive focus on the pathways of rational thought involved in risk responses is misguided. A refrain throughout this chapter will be that rather than the more rational, cognitive processes currently given prominence, a more emotive dimension lies at the core of the response to danger. This is not to say that the former is not involved: dual-process theories suggest that 'thought' might comprise both a rapid, emotional dimension and a stream devoted to slower, more deliberative cognitive processing (Epstein 1994). In recent times, dual-process theories have increasingly found voice in the work of psychologists working on risk perception, who conceive of responses to potential dangers as operating along rapid 'feeling' and slower 'analysis' dimensions (Slovic et al. 2004; Slovic and Peters 2006). Key theorists of the 'risk perception' model are now redressing the relegation of emotion by giving primacy to 'the feeling of risk', as is the title of Slovic's (2010) more recent anthology, in contrast to his earlier 'The perception of risk' (2000). However, the psychology that finds its way into other fields and applied contexts often remains mired in older models of risk perception with their highly cognitivist, rational focus on probability assessment. Further, it must be noted that the 'risk as feeling' trend within the psychology of risk perception revolves largely around 'affect', which Slovic et al. (2005) define as a "faint whisper of emotion" (p. 353) demarcating the goodness or badness of a stimulus. Affect is experienced as a feeling, either consciously or non-consciously and denotes a positive or negative aspect of the stimulus. Slovic et al. (2005) state that their use of the concept does not refer to the more 'visceral emotion of fear' that people may experience regarding a potential danger. Much of Leiserowitz's work (e.g. Leiserowitz 2006; Smith and Leiserowitz 2012) on climate change affect typifies this trend.

A further difficulty with the way in which risk perception research has been pursued lies in the somewhat dismissive orientation it adopts to lay knowledge. Much risk 'perception' research could more accurately be characterised as research on risk 'misperception'. The risk perception research agenda often invokes the notion of a fundamental division between the knowledge held by experts, which is assumed to be objective, correct and authoritative, and that held by laypeople, which is subjective, irrational and often erroneous. Scientific evaluations of risks, from food scares to climate change to natural disasters, are rarely wholly unanimous or immutable; Beck (2006) contends that it is scientists whose "findings often contradict each other, who change their minds so fundamentally that what was judged 'safe' to swallow today, may be a 'cancer risk' in two years' time." (p. 345). Indeed, fallibility is a cornerstone of scientific method and philosophy. Despite this, it is laypeople whose (mis)judgements and biases are emphasised, with deviations between lay beliefs and current expert consensus invariably attributed to errors on the part of the lay thinker. In particular, the risk perception field examines the numerical probabilities people offer in estimating their chances of being affected by a particular risk. These are then compared to scientific estimates to identify how the public distort risk information, typified in the frequent laments about the public's amplification of certain hazards that experts deem low-risk, such as air travel, and underestimation of those, including car travel, which they 'should' rate as high risk.

The focus of the risk perception field on perceptual and cognitive error in apprehending expert-defined risks often implicitly endorses a 'deficit model' of public responses to scientific information. The deficit model approach to lay knowledge assumes that the public are ignorant of scientific 'facts' and seeks to rectify what is seen as an information- or knowledge-gap between science and its lay recipients (Hilgartner 1990). Deficit model approaches have been criticised both for reifying expert perspectives as objective truth and for propagating a view of a single, context-independent rationality, thereby precluding recognition of alternative forms of rationality that operate within local contexts. For example, Graham's (1993) study of attitudes to smoking risk among socio-economically disadvantaged single mothers showed that women were aware of smoking's health risks, but consciously judged that these were outweighed by the benefits they derived from smoking, in providing a form of stress relief in harsh social circumstances. Risk evaluations that deviate from expert advice are therefore not necessarily irrational or uninformed, but rather can reflect the contingencies with which people live. Within theorisation of science's role in society, the deficit model has now fallen from favour, with a focus on public 'understanding' of science shifting to public 'engagement' with science (Bauer 2009). The shift to public engagement with science leaves open alternative ways of theorising public responses to risk, which do not privilege one particular form of thinking and that examine the texture and consequences of lay thought without an agenda of judging whether it is 'right' or 'wrong'.

2.4 Evidence for 'Risk Perception' Driving Action?

A key ambition of those who use risk perception research in applied contexts is the development of effective means of predicting (and, it is hoped, thereby understanding and changing) people's risk-related behaviour. Thus a question arises concerning its explanatory power in the area of behaviour: are people's danger-related behaviours driven by the cognitive components of responses to danger, or by other elements that are based in the emotional and socio-cultural experience of risk?

Efforts to change people's risk behaviour have traditionally relied on 'public education' campaigns, assuming that if people are aware of a risk and know what to do, they are likely to act accordingly. This assumption is rooted in the rationalist philosophy that has dominated the psychology of risk since the mid-twentieth century, typified in the names of prominent models of the thought-action relationship such as the

'Theory of Reasoned Action' and 'Theory of Planned Behaviour' (Ajzen 1991; Fishbein and Ajzen 1975). The limits of rationalist understandings of risk-related behaviour have gradually become evident (see Joffe 1996). In the area of adjusting to earthquake threat, for example, research generally concludes that awareness of earthquake risk and some of the variables contained in the cognitive models are only weakly related to undertaking protective behaviour (Solberg et al. 2010). Solberg et al.'s (2010) review highlights the importance of factors with emotional and/or socio-cultural foundations, such as trust, social identity, attributions of responsibility and fatalism in informing earthquake preparedness. Extensive research in the area of emerging infectious diseases has also corroborated the importance of trust and the emotional experience of worry in influencing the actions people take to reduce their risk of infection (Bish and Michie 2010; Rubin et al. 2010). Buttressed by such findings, recognition of the limits of purely cognitive approaches appears to be dissipating further into the risk perception field: 25 years after research began on the notion of 'optimistic bias' (Weinstein 1978), a 'cognitive distortion' whereby people see themselves as less vulnerable to dangers than their peers, the psychologist who pioneered the concept has begun to change his take on what drives behaviour, shifting from the purely cognitive to the more emotive realm: "considerable evidence exists demonstrating that worry (a poorly defined concept containing elements of both emotion and attention) [...] provides predictive power beyond that provided by judgements of likelihood and severity" (Weinstein 2003, p. 48). Indeed, the very phenomenon of optimistic bias may be fundamentally emotionally driven, reflecting a defensive response to fear rather than 'cold' cognitive error (Joffe 1999).

Thus, it seems that faulty judgements are not the key or sole predictors of people's responses to risks. With an understanding of 'risk' as 'potential danger' (Douglas 1992), or as Beck (2006) puts it, 'anticipated catastrophe', what comes to the fore is emotion in relation to threat. Theorists such as Beck suggest that from a technical perspective, part of the work of modernity is to set in motion mechanisms to control the incalculable and unknowable – from vaccination to early detection systems, from retrofitting to the science of risk perception. This machinery of control does not always function as intended or desired, and humans must develop other means of managing the emotional challenges that arise in times of crisis.

2.5 A More Fitting Theory of People's Apprehension of Dangers: From 'Risk Perception' to 'Risk Representation'

When a focus on the accuracy of public 'understandings' or 'perceptions' of risks is replaced by a concern with public 'engagement' with potential dangers, respect for local knowledge becomes central. This chapter posits a theory that takes as its subject the common-sense of local knowledge: social representations theory (SRT), a social psychological theory that examines lay uptake of societal phenomena (Moscovici 1961/2008). Social representations are widespread 'ensembles of thoughts and feelings' about a given phenomenon akin to a 'lay theory' or 'lay explanation'. This 'ensemble of thoughts and feelings' is expressed by individuals in their conversation, behaviour and internal understandings (Wagner et al. 1999) but underpinned by socio-cultural, historical and group-specific ideologies. Thus, with SRT the theoretical emphasis shifts from the individual mind as a processor of risk information to people and their ideas as products of their histories, social groups and cultures. Furthermore, the theory contends that people "generate representations not so much by a reasoning or computing process as by a process of communicating" (Moscovici 1984, p. 951). This is particularly relevant to representations of risks such as earthquakes, climate change and flooding, which do not merely face individuals but neighbourhoods, communities and countries. Representations of these phenomena are forged in the ongoing communication that occurs between people, and a considerable portion of the representations of these threats is therefore socially shared. The aim of SRT is to document this shared portion of knowledge about particular phenomena. The foci thus become exploration of the substantive content of what people say about threats (as opposed to, for example, numerical probability estimates) and examination of widespread, culturally-based representations of the threat (as opposed to individual-level variables).

SRT holds that in the course of communication about a new risk, two processes occur: 'anchoring', by which existing ideas and concepts are imposed upon the new information, and 'objectification', where the new information is assimilated into people's knowledge by means of familiar symbols, metaphors and images. For example, Japanese mythology represents earthquakes as a consequence of the movement of giant catfish (*Namazu*) who live underground (Yamori 2013). Official Japanese earthquake information communications today continue to recruit images of catfish: despite the catfish explanation of earthquakes not being literally believed, the historical symbol has persisted and effortlessly connotes 'earthquake' to Japanese people.

An important facet of the application of social representations theory to studying responses to risk relates to its incorporation of the emotive dimension of knowledge. For example, Höijer's (2010) analysis of coverage of climate change in Swedish newspapers pinpoints the importance of the emotions of fear, hope, guilt, compassion and nostalgia in constituting the social representations that evolve. Moscovici (1961/2008) suggests that social representations emerge in response to a threat to a group's identity or way of life, and that a central function of social representation is to defend against such threat by 'making the unfamiliar familiar'. Psychological efforts to manage emotions like anxiety can therefore shape the representations of risk that evolve. For example, Joffe's (1999) research documents how social representations of AIDS in the 1990s developed along a pattern of 'not me, the other is to blame', with a variety of different groups downgrading their own vulnerability to HIV/AIDS and focusing on the inflated risk they attributed to dissimilar 'others', such as gay men or people from other countries. This tendency to distance the self from worrying phenomena by projecting risk onto 'the other' permeates responses to a wide variety of risks, including a range of infectious diseases (Joffe 2011), climate change (Smith and Joffe 2013) and earthquakes (Joffe et al. in press).

A strength of the social representations framework as a theory of public engagement lies in its ability to see different publics' accounts of a given danger as forms of knowledge in their own right rather than as deficient forms of scientific knowledge (Duveen 2008). SRT examines the contents of people's risk-related thinking without reference to the 'reality' of the risk; the concern is not with the accuracy of the representation but with the meanings that people attach to a given risk and the consequences of these meanings for themselves, for others and for society. These 'common-sense' notions are no less important than people's more scientifically-informed knowledge in shaping their responses to the risks they encounter. On a practical note, this is an important consideration for policy makers seeking to understand how people's danger-related thinking 'works' and might therefore be challenged and changed: focusing purely on the accuracy of people's scientific knowledge means that other dimensions of their risk-related thought, feeling and behaviour remain undocumented. More naturalistic studies that focus on where people *are* rather than where they *should* be have clear advantages for developing interventions. As Fishbein and Cappella (2006, p. S4) state, "one must understand the behavior from the perspective of the population for whom the interventions are being developed. Once understood this way, these beliefs can serve as the basis for messages and other interventions". A further advantage that SRT offers for policy concerns relates to its concentration on the communicative aspects of social representations. It interrogates the role played by messages that circulate in the public sphere, particularly within the mass media, in shaping risk representations (e.g. Washer et al. 2008). Such analyses can indicate how change can be effected in threat-related behaviour through changes in the communications people encounter.

To illustrate the usefulness of adopting a view of responses to risk as largely emotional and socio-cultural phenomena, this chapter introduces an empirical study of how people living in the cities of Seattle (USA), Izmir (Turkey) and Osaka (Japan) – all areas that are highly seismic but that have not recently been directly affected by damaging earthquakes – represent earthquakes.

2.6 A New Way of Studying Responses to Earthquake Risk: The EPICentre Risk Representation Project

Investigation of naturalistic representations of a given threat involves a departure from conventional psychological methods of experimentation and survey research. Survey and experimental methods require that a researcher predefine the variables of interest, leaving little space for the emergence of findings that have not been anticipated. This is particularly problematic when there is a cultural gap between the researcher and participant, as the cross-cultural equivalence of concepts cannot be assumed. Furthermore, surveys, in particular, tap only consciously available cognitions. If people's risk-related ideas are informed by values, images and emotions that are not necessarily conscious, this material will not become visible in self-reported questionnaire responses. An additional issue relates to the extent to which the psychological response to risk can be captured using purely quantitative measures, or whether important facets of the texture of thought are lost when it is translated into a series of numbers.

These concerns with the validity of conventional methods of investigating human responses to risk stimulated a researcher working within UCL's Earthquake and People Interaction Centre (EPICentre) to develop a new research technique, combining features of free-association with qualitative interview methodology (Joffe 2011; Joffe and Elsey, under review). The free associative interview technique involves presenting participants with an empty four-box grid and asking them to write or draw the first four words, feelings or images that come to mind when exposed to a certain prompt – in this case, the word 'earthquake'. The researcher then conducts an interview that is structured around these free-associations, asking the respondent to expand upon the associations they produced and asking follow-up questions to prompt the respondent to elaborate further. The interviewer avoids posing questions about areas or ideas that have not been spontaneously volunteered by the respondent, thereby ensuring that the researcher's preconceptions have minimal influence on the material gathered. This provides a greater degree of confidence that the data reflect the individual's naturalistically-occurring thought patterns and tap the stored, automatic links made to the risk in question. Furthermore, free associations reveal people's most salient associations to the phenomenon, which Hollway and Jefferson (2000) suggest are primarily emotive. Thus the method provides an entry-point into the emotional substructure of people's responses to risk. The interviews are recorded, transcribed verbatim, and analysed using qualitative analytic procedures such as thematic analysis in order to identify any patterns in individuals' responses to the risk. The data that emerge from this method are distinctive in the extent to which they reveal latent emotional and socio-cultural drivers of responses to risk.

EPICentre's project involved a cross-cultural application of this method to interview demographically-matched samples of 48 individuals in each of three earthquake-prone cities – Seattle, Izmir and Osaka. The results of this study, presented in Joffe et al. (in press), contradict the conventional assumption that action to reduce one's risk results from perceiving a risk as likely and severe. Questionnaire data collected alongside the interviews (see Rossetto et al. 2011) indicated that despite universally high awareness of earthquake threat, participants in all three cities did little to prepare for earthquakes or mitigate their risk. Out of a list of 19 risk-reducing actions, participants in all locations reported having done (on average) less than half, though the Americans had adopted a significantly greater number of actions than either the Turkish or Japanese participants.

Qualitative analysis of the interview data threw light on people's failure to act to reduce their earthquake risk. Analysis indicated that important contributors towards whether people acted or not lay in their emotional responses to earthquake threat. While the Japanese and Turkish respondents associated earthquakes with intense panic, fear and anxiety, American respondents displayed a much more moderate level of concern. Indeed, positive emotional responses to earthquakes were not uncommon in the US sample, with earthquakes provoking emotions such as awe, exhilaration and fascination. The Turkish interviews were notable for the intensity of emotions related to anger, blame and distrust of state and construction authorities, whose negligence and corruption were positioned as the cause of earthquake damage. While US and Japanese participants distanced themselves from threat by favourably comparing themselves to other places that they saw as subject to a greater earthquake risk, Turkish participants sharply differed from this pattern: they also compared themselves with other countries, but with the outcome of reinforcing their sense that *they themselves* were extremely vulnerable to earthquake damage.

These emotional responses to earthquake threat interacted with a cluster of feelings relating to a sense of agency and control in relation to earthquakes. Almost all participants demonstrated at some point in the interview that they were aware that actions were available by which they could reduce their risk from earthquakes. However, alongside this awareness lay strong trends towards fatalism in all three cities, with respondents asserting that human action is pointless in the face of earthquake threat. These fatalistic tendencies were clearly culturally grounded and were rooted in different belief-systems in the three countries. US and Japanese fatalism largely followed from cultural representations of earthquakes as acts of nature, which is a sphere immune from human influence. The fatalism present in the Turkish data, on the other hand, drew on a religious framing of earthquakes as acts of divine power, as well as an emphasis on vulnerability caused by institutional corruption.

Framing the failure to act within the emotional and socio-cultural contexts revealed in the interviews furnishes important insights into the risk experience of citizens of earthquake-prone cities. The data suggest that emotional and socio-cultural variables exert an appreciable influence on people's risk-relevant behaviour. The intense emotions experienced by the Turkish and Japanese participants, in combination with cultural currents that represented earthquakes as insurmountable forces, seem to have overwhelmed people and paralysed them from action. The US participants displayed a more muted emotional response to earthquakes and also reported the greatest levels of protective action. A more moderate level of concern, rather than high anxiety, may be more facilitative of earthquake-protective behaviour. Nevertheless, even the US participants performed less than half of the recommended practices, with action possibly undermined by cultural representations of earthquake threat as impervious to human action.

2.7 Concluding Comments

This chapter has polarised the more cognitive and the more emotive and cultural approaches to public engagement with risks. It has leant towards championing the latter since this rectifies an imbalance in the literature. However, the facets work in tandem. The method adopted in the EPICentre project provides the groundwork for establishing a valid model for risk-related thought, feeling and ultimately behaviour. Beyond its qualitative exploration of free associations, it also explored some of the more traditional variables found in 'risk perception' studies to gain an holistic picture of how people engaged with earthquake risk.

When 'risk perception' is approached in this way, we see that the public's assessments of potential dangers do not differ from those of experts purely because they cannot do the sums. Rather, they care about different things: people's representations of risk are not solely based on evaluations of probability and severity, but incorporate a wide range of relevant information about the personal, community and societal contexts in which the disaster would occur. Izmir's residents' failure to act is not a consequence of ignorance or inaccurate judgement, for example, but of their representation of the structures in which they live as badly built. High levels of structural vulnerability are represented as resulting from corruption in the building industry and among those who (fail to) regulate it. The sense of living in unsafe structures undermines people's self-efficacy as they feel that their actions will be pointless since their buildings are not built to withstand earthquakes. A decision not to devote resources to mitigative activities is understandable in this light. The shift from traditional 'risk perception' to 'representations of potential catastrophes' acknowledges the contextual and multi-faceted nature of lay knowledge, and examines it on its own terms without an agenda of determining whether it is deficient or adequate with reference to an ideal 'expert' position.

This respectful orientation to lay knowledge provides a clear advantage when incorporating research into policy implementation or behaviour change programmes. Though methodologically convenient, squeezing lay experience into a form that it does not take naturalistically means that resultant policy initiatives draw on a contorted picture of ordinary thought and behaviour, leaving their implementation vulnerable to unanticipated reactions. A social representations approach starts with where people are rather than where they 'should' be, thereby furnishing a valid and comprehensive depiction of lay experience. This map of the contours of social life can be used by policy initiatives to plan an implementation strategy that promotes public engagement and support.

The EPICentre study provides a counter-balance to predominantly cognitive perspectives on risk perception. Social representations are highly dynamic: they are "networks of interacting concepts and images whose contents evolve continuously over time and space" (Moscovici 1988, p. 220). Thus social representations can change. SRT's focus on the dependence of internal representation on the external socio-cultural environment brings to the fore the role played by the media and other societal messages in forging representations. SRT suggests that the interaction between these messages and individual minds will play a strong role in framing people's behavioural choices. SRT studies can thereby indicate the types of messages that may encourage or undermine taking action against various risks.

Since SRT research shows that the network of concepts, images and associations that define a social representation are specific to certain groups, a one-size-fits-all communication strategy is unlikely to impact upon different sectors of a society in the same way. For example, the health sphere has seen much interest in the effectiveness of campaigns that recruit emotional responses such as disgust (and indeed surprise, humour and fear) to bring about change in health-related behaviours. A key contingency in such campaigns is that different groups respond differently to these emotively-tinged messages; for example, one study found that disgust-inducing messages significantly increased the practice of hand-washing in men but not in women (Judah et al. 2009). The take-home message is that *tailoring* risk messages to particular audiences is crucial. Such tailoring can only be achieved once the dynamics of each group's representation of a particular danger are understood.

In terms of the broader context, contemporary societies appear to organise around a science of prediction, both in terms of predicting the dangers themselves and people's responses to them. Yet prediction is at its least possible regarding potential catastrophes. Unknowable risks hide behind a façade of calculability in the risk society (Beck 2006). Stepping outside of the conventional conceptualisation of lay apprehension of potential danger that is seen in mainstream risk perception research brings emotion and contemporary cultural thinking centre-stage. It is these facets of the response to potential danger, alongside more cognitive aspects of risk perception, that need to be understood in order to tackle the obdurate problem of low levels of preparatory behaviours in the face of risks.

References

Ajzen I (1991) The theory of planned behavior. Organ Behav Hum Decis Process 50(2):179–211

Bauer MW (2009) The evolution of public understanding of science – discourse and comparative evidence. Sci Technol Soc 14(2):221–240

Beck U (1992) Risk society: towards a new modernity. Sage, London

Beck U (1996) Risk society and the provident state. In: Lash S, Szerszynski B, Wynne B (eds) Risk, environment and modernity: towards a new ecology. Sage, London, pp 27–43

Beck U (2006) Living in the world risk society. Econ Soc 35(3):329–345

Bernstein PL (1998) Against the gods: the remarkable story of risk. Wiley, New York

Bish A, Michie S (2010) Demographic and attitudinal determinants of protective behaviours during a pandemic: a review. Br J Health Psychol 15(4):797–824

Douglas M (1992) Risk and blame: essays in cultural theory. Routledge, London

Douglas M, Wildavsky A (1982) Risk and culture: an essay on the selection of technical and environmental dangers. University of California Press, Berkeley

Duveen G (2008) Social actors and social groups: a return to heterogeneity in social psychology. J Theory Soc Behav 38(4):369–374

Epstein S (1994) Integration of the cognitive and the psychodynamic unconscious. Am Psychol 49(8):709–724

Fishbein M, Ajzen I (1975) Belief, attitude, intention, and behavior: an introduction to theory and research. Addison-Wesley, Reading

Fishbein M, Cappella JN (2006) The role of theory in developing effective health communications. J Commun 56(s1):S1–S17

Graham H (1993) When life's a drag: women, smoking and disadvantage. HMSO, London

Hilgartner S (1990) The dominant view of popularisation: conceptual problems, political uses. Soc Stud Sci 20(3):519–539

Höijer B (2010) Emotional anchoring and objectification in the media reporting on climate change. Public Underst Sci 19(6):717–731

Hollway W, Jefferson T (2000) Doing qualitative research differently: free association, narrative and the interview method. Sage, London

Hough S (2010) Predicting the unpredictable: the tumultuous science of earthquake prediction. Princeton University Press, Princeton

Joffe H (1996) AIDS research and prevention: a social representational approach. Br J Med Psychol 69(3):169–190

Joffe H (1999) Risk and 'the other'. Cambridge University Press, Cambridge

Joffe H (2003) Risk: from perception to social representation. Br J Soc Psychol 42(1):55–73

Joffe H (2011) Public apprehension of emerging infectious diseases: are changes afoot? Public Underst Sci 20(4):446–460
Joffe H (2013) The L'Aquila quake decision and the stigma against science. Risk Manage 60(1):6–7
Joffe H, Elsey J (under review) The free associative interview technique
Joffe H, Haarhoff G (2002) Representations of far-flung illnesses: the case of Ebola in Britain. Soc Sci Med 54(6):955–969
Joffe H, Rossetto T, Bradley C, O'Connor C (submitted) Stigma in science: the case of earthquake prediction
Joffe H, Rossetto T, Solberg C, O'Connor C (in press) Social representations of earthquakes: a study of people living in three highly seismic areas. Earthquake Spectra
Judah G, Aunger R, Schmidt W-P, Michie S, Granger S, Curtis V (2009) Experimental pretesting of hand-washing interventions in a natural setting. Am J Public Health 99(S2):S405–S411
Kahan DM, Jenkins-Smith H, Braman D (2010) Cultural cognition of scientific consensus. J Risk Res 14(2):147–174
Leiserowitz T (2006) Climate change risk perception and policy preferences: the role of affect, imagery and values. Clim Chang 77(1–2):45–72
Lerner JS, Gonzalez RM, Small DA, Fischhoff B (2003) Effects of fear and anger on perceived risks of terrorism: a national field experiment. Psychol Sci 14(2):144–150
Moscovici S (1961/2008) Psychoanalysis: its image and its public. Polity Press, Cambridge
Moscovici S (1984) The myth of the lonely paradigm: a rejoinder. Soc Res 51(4):939–969
Moscovici S (1988) Notes towards a description of social representations. Eur J Soc Psychol 18(3):211–250
Nishikawa S (2011) Japan's preparedness and the great earthquake and tsunami. In: Funabashi Y, Takenaka H (eds) Lessons from the disaster. The Japan Times, Tokyo, pp 17–47
Peters EM, Burraston B, Mertz CK (2004) An emotion-based model of risk perception and stigma susceptibility: cognitive appraisals of emotion, affective reactivity, worldviews, and risk perceptions in the generation of technological stigma. Risk Anal 24(5):1349–1367
Rossetto T, Joffe H, Solberg C (2011) A different view on human vulnerability to earthquakes: lessons from risk perception studies. In: Spence RJ, So E, Scawthorn C (eds) Human casualties in earthquakes: progress in modelling and mitigation. Springer, Dordrecht, pp 291–304
Rubin GJ, Potts HWW, Michie S (2010) The impact of communications about swine flu (influenza A H1N1v) on public responses to the outbreak: results from 36 national telephone surveys in the UK. Health Technol Assess 14(34):183–266
Sarewitz D (2010) World view: tomorrow never knows. Nature 463:24
Scoones I, Foster P (2010) Unpacking the international response to avian influenza: actors, networks and narratives. In: Scoones I (ed) Avian influenza, science, policy and politics. Earthscan, London, pp 19–64
Sjöberg L (2007) Emotions and risk perception. Risk Manage 9(4):223–237
Slovic P (ed) (2000) The perception of risk. Earthscan, London
Slovic P (ed) (2010) The feeling of risk. Earthscan, London
Slovic P, Peters E (2006) Risk perception and affect. Curr Dir Psychol Sci 15(6):322–325
Slovic P, Finucane ML, Peters E, MacGregor DG (2004) Risk as analysis and risk as feelings: some thoughts about affect, reason, risk, and rationality. Risk Anal 24(2):311–322
Slovic P, Peters E, Finucane ML, MacGregor DG (2005) Affect, risk, and decision making. Health Psychol 24(4):S35–S40
Smith N, Joffe H (2013) How the public engages with global warming: a social representations approach. Public Underst Sci 22:16–32
Smith N, Leiserowitz A (2012) The rise of global warming scepticism: exploring affective image associations in the United States over time. Risk Anal 32(6):1021–1032
Solberg C, Rossetto T, Joffe H (2010) The social psychology of seismic adjustment. Nat Hazard Earth Syst Sci 10(8):1663–1677

Wagner W, Duveen G, Farr R, Jovchelovitch S, Lorenzi-Cioldi F, Markova I (1999) Theory and method of social representations. Asian J Soc Psychol 2(1):95–125

Washer P (2004) Representations of SARS in the British newspapers. Soc Sci Med 59(12):2561–2571

Washer P (2006) Representations of mad cow disease. Soc Sci Med 62(2):457–466

Washer P, Joffe H (2006) The "hospital superbug": social representations of MRSA. Soc Sci Med 63(8):2141–2152

Washer P, Joffe H, Solberg C (2008) Audience readings of media messages about MRSA. J Hosp Infect 70(1):42–47

Weinstein ND (1978) Cognitive processes and information seeking concerning an environmental health threat. J Human Stress 4(1):32–41

Weinstein ND (2003) Exploring the links between risk perceptions and preventive health behavior. In: Suls J, Wallston KA (eds) Social psychological foundations of health and illness. Blackwell Publishing, Oxford, pp 22–53

World Health Organisation (2012) The Great East Japan earthquake: a story of devastating natural disaster, a tale of human compassion. WHO Press, Geneva

Yamori K (2013) A historical overview of social representation of earthquake risk in Japan: Fatalism, social reform, scientific control and collaborative risk management. In: Joffe H, Rossetto T, Adams J (eds) Cities at risk: living with perils in the twenty-first century. Springer, Dordrecht, pp 73–91

Chapter 3
Risk Compensation in Cities at Risk

John Adams

Abstract Cities face many risks. This book focuses mainly on natural hazards. This chapter explores what perceptions of, and responses to, natural hazards have in common with society's way of coping with risk more generally. It proffers four *framing devices*. First a categorisation of risk: is it directly perceptible, perceptible only with the help of scientific instruments and statistical surveys, or is it 'virtual' – perceptible only with the assistance of unquantifiable imagination? Second, the *risk thermostat* introduces the concept of *risk compensation*: this presents the idea that risk perceptions influence behaviour in predictable ways. Third, *Cultural Theory* introduces a framework for organizing the biases that one encounters when trying to understand diverse responses to what appear to be statistically similar risks. Finally, the significance of *voluntary versus imposed* risks: why are there such enormous differences in responses to statistically similar risks?

3.1 The Risks We Face

A few years ago I was in Amsterdam for a conference on road safety. I was waiting to meet someone in the lobby of a modern hotel in the outskirts of the city. I felt safe. No one in the lobby looked threatening: no robbers, terrorists or psychopaths. The building felt safe. I was sitting next to a massive concrete pillar that appeared to my non-expert eye more than sufficient to hold the building up. But then I tipped back in my chair to look at the ceiling – I was bored, the person for whom I was waiting was late. On the concrete pillar some feet above my head was a wavy line inscribed with the words 'sea level'.

J. Adams (✉)
Department of Geography, University College London,
Gower Street, London WC1E 6BT, UK
e-mail: john.adams@ucl.ac.uk

H. Joffe et al. (eds.), *Cities at Risk: Living with Perils in the 21st Century*,
Advances in Natural and Technological Hazards Research 33,
DOI 10.1007/978-94-007-6184-1_3, © Springer Science+Business Media Dordrecht 2013

Suddenly, if briefly, I was alarmed. As a student of risk I am practiced at imagining disasters: what if the flood defences were breached by a storm or, perhaps, by accident or malign intent? I had heard about the great flood of 1953 in which over 1,800 people in the Netherlands died. But I knew, rather vaguely, that much had been done since then to strengthen the flood defences and, much more important in calming my nerves, was the sight of normal life going on all around me. I relaxed.

Subsequent inquiry revealed that since 1953 a great deal had been done. Twenty days after the flood the Delta Commission was inaugurated. In 1959 a plan to build extra flood defences, the Delta Project, was enshrined in the Delta Law. It was a scheme of such ambition that the American Society of Civil Engineers proclaimed it one of the Seven Wonders of the Modern World (Deltawerken Online 2011).

About 20% of the area of the Netherlands (including Amsterdam and large parts of Rotterdam) is below sea level (Schenau et al. 2009). About two thirds of the population–more than 10 million people–live in areas vulnerable to flooding (van Alphen 2009). For such large numbers, a total evacuation in the face of a flood threat is not a realistic option. So, how to defend them? And at what cost?

According to one account, the Commission initially set the 'acceptable risk' of catastrophic flooding at once every 125,000 years (Jonkman et al. 2005). This, however, was deemed too expensive and the acceptable risk around which it was ultimately planned was increased to once every 10,000 years. How were such numbers arrived at? The Delta Works Wikipedia entry claims that:

> The cost of flooding is assessed using a statistical model involving damage to property, lost production and given amount per human life lost. For the purpose of this model a human life is valued at €2.2 million (2008 data). The chances of a significant flood within the given area are calculated. This is done using data from a purpose-built flood simulation lab as well as empirical statistical data regarding water wave properties and distribution. Storm behaviour and spring tide distribution are also taken into account. (Wikipedia 2012)

What kind of risk is the Netherlands dealing with? The Venn diagram of Fig. 3.1 identifies three different types of risk with which we all, as risk managers, wrestle. The diagram could, of course, be overlaid with a multitude of other circles – health risk, emotional risk, enterprise risk, financial risk, reputational risk, value at risk, fraud risk, political risk, military risk, security risk, traffic risk, environmental risk – but the three in the diagram capture essential attributes of all the others.

Some risks are visible to the naked eye, that is 'directly perceived'. We manage them using judgment; we do not undertake a formal probabilistic risk assessment before crossing the road. Some combination of instinct, intuition and experience usually gets us safely to the other side.

Others are perceptible only to those armed with the tools of science – microscopes, telescopes, scanners and other measuring devices, surveys, and the data they produce. This is the realm of quantified risk management. In this realm uncertainty is qualified by probability.

Finally, 'virtual risk' may or may not be 'real' – scientists disagree – but they have real consequences. For some the uncertainty is liberating; if science cannot settle the issue they feel free to argue from their beliefs, convictions, prejudices or superstitions. Here we are thrown back, as in the first circle, on judgments that cannot be objectively validated.

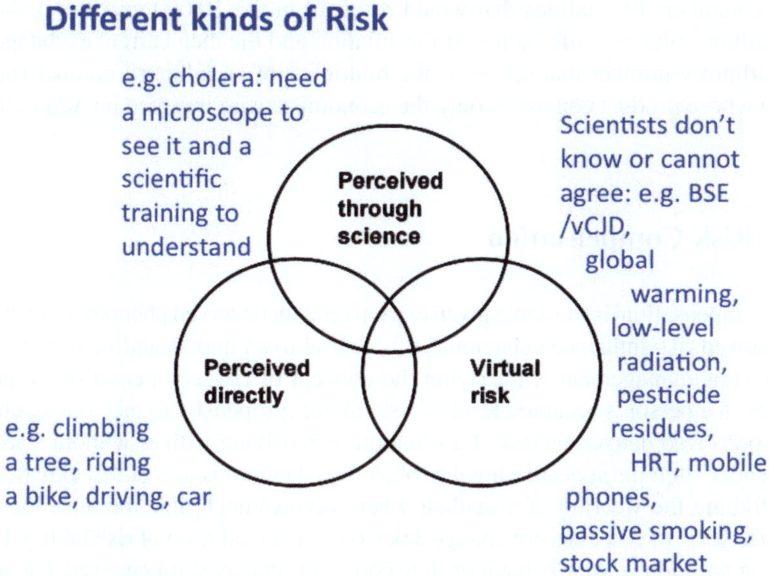

Fig. 3.1 Different kinds of risk

The Dutch history of flooding illustrates all three types. The 1953 flood was far from the worst. The Grote Mandrenke flood of 1362 is reported to have killed over 25,000 people and the 1,530 St. Felix's Flood over 100,000 (Moss 2011; Borthwick 2011). The individual or local response to the directly perceived risk is guided by centuries of experience and mythology. My favourite example I picked up at conference in Delft on the risks of nuclear energy. The conference being in the Netherlands, the discussion amongst the assembled risk experts soon turned from nuclear risks to flood risks. One participant described an intriguing practice from an earlier century: when the closely monitored flood reached a certain height the men of the village, armed with picks and shovels, would row across the river and breach the dyke on the other side. Mythology or not it sounded a plausible, if morally dubious, response to a directly perceptible risk.

Much of modern Dutch flood-control practice can be assigned to the second circle of Fig. 3.1. Little responsibility is now left to the individual citizen, or village. The scale of the threat is seen as so great that a collective response is required. The risk is perceived through science and managed by the state; and, as illustrated above through the Delta Commission, whose management is guided by large databases and computer models.

However, the greater part of current flood control planning takes place in the virtual risk circle of Fig. 3.1. How might the managers decide between defences that protect against a 125,000 year flood and a 10,000 year flood? The €2.2 million value placed by the modellers on a human life in 2008 is highly contentious, as is whatever discount rate that they might use to project it thousands of years into the future. Although numbers are paraded in justification of the proposals, they are wild speculation in the case of numbers of fatalities that would result from the 10,000 year flood. And the €2.2 million value of a life (adjusted for inflation and the then current exchange rate) is an arbitrary number that serves as the multiplier of an arbitrary number (number killed) whose product convinces only the economists who invented it (Adams 1974).

3.2 Risk Compensation

'Risk compensation' is the name given to a universally observed phenomenon: changes in perceived risk influence behaviour.[1] It is central to an understanding of the process of risk (mis)management. Underlying the concept of risk compensation is the idea that, given a person's accepted level of risk taking (propensity to take risks), changes in the perceived dangerousness of a situation (e.g. driving with or without a seatbelt) may render attempts at decreasing the 'objective' dangerousness of that situation (e.g. by enforcing the wearing of a seatbelt when driving) negligible, because the newly modified level of risk does not change a person's accepted level of risk taking. Hence, a person will adjust their behaviour in accord with changes in perceived risk, essentially maintaining their level of danger. As a concrete example, laws enforcing the wearing of a seatbelt may make the event of a car *crash* safer, yet the number of road deaths may remain constant, or even increase. In terms of risk compensation, this can be explained because the enforcement neglects the changes in driving behaviour engendered by the sense of safety that wearing a seatbelt might produce. Figure 3.2 proffers the essence of the process. It describes the 'risk thermostat' that every one of us employs in the face of uncertainty. 'Propensity' in this diagram represents the setting of the thermostat. Some thermostats are set high, others low. I have yet to meet anyone with a zero setting; life would be unutterably boring.

Propensity in this model leads to risk taking behaviour that leads, by definition, to accidents. To take a risk is to do something that carries with it a probability of an adverse outcome. Through surviving accidents and learning from them, or seeing them on television, or being warned by mother, we acquire our perception of safety and danger. The model postulates that when propensity and perception get out of balance we behave in a way that seeks to restore the balance. Why do we take risks? There are rewards, and the model proposes that the magnitude of the reward influences propensity.

Unlike individual risk managers crossing roads or riding bicycles, most institutional risk managers, outside the offices of venture capitalists, hedge funds managers and sub-prime mortgage brokers, are dedicated to the prevention of bad things happening. They are focused on the bottom loop of Fig. 3.2. They are risk-averse. But as people or societies become more risk-averse they do not necessarily become safer or more resilient.

---

[1] The term was coined by Canadian psychologist Gerald Wilde (1976) and elaborated in Adams (1985) and Adams (1995).

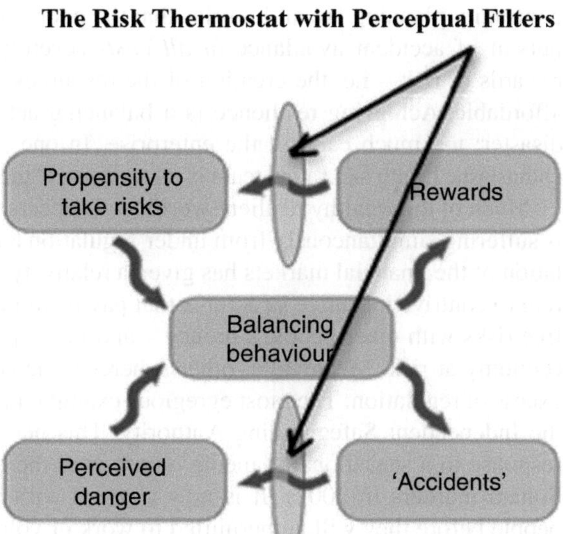

Fig. 3.2 The risk compensation process: The risk thermostat with perceptual filters (see Sect. 3.4 for discussion of the perceptual filters)

3.3 Risk and Resilience

Increased resilience is the objective of all disaster planning, whether individual or societal. It is a relative quality. There are no agreed units by which it can be measured[2] but some people, and societies, have more of it than others. The ability to prevent bad things happening, to mitigate their consequences and to speed recovery when they do, is not equitably distributed.

It is also limited. Ultimately the pursuit of it ends in failure. Empires collapse, companies go out of business, everyone dies. On a geological time scale tectonic plates shift, ice ages come and go, asteroids impact, the sun goes cold. On a human time scale tsunamis, earthquakes, credit crunches, diseases and simple traffic accidents[3] can overwhelm the most resilient of individuals. But still we strive to prevent bad things happening, mitigate their consequences and speed recovery.

Resilience requires command over resources. Building flood defences and earthquake resistant buildings, accident and emergency services, and post-disaster

[2] Costs and benefits measured in various currencies are sometimes suggested but are not helpful metrics for low frequency high impact events such as 125,000-year, or even 10,000-year, floods. Leaving aside the intractable problems of discount rates, fluctuating currencies, and agreeing the current cash value of human lives, such time spans pre-date all known currencies and exceed the survival prospects of any existing currencies.

[3] Deaths caused by traffic accidents and terrorism illustrate the measurement problem. Resilience is often a term encountered in discussions about the efficacy of anti-terrorism measures, but never in my experience in discussions about the efficacy of measures to curb road accidents, which kill far more people.

continuity planning are all luxuries that the poor cannot afford. The single-minded pursuit of accident avoidance *at all costs* severely constrains the pursuit of the rewards of risk – i.e. the creation of the resources that ultimately make resilience affordable. Achieving resilience is a balancing act. Too little caution can lead to disaster; too much can kill the enterprise. In one company I know, the (overly?) enthusiastic health and safety team is referred to as "the sales prevention department."

Much of the wealthy resilient world now appears to be becoming less resilient. It is suffering simultaneously from under-regulation and over-regulation. The deregulation of the financial markets has given a relatively small number of financiers free rein to contrive incentive structures that pay them fabulous rewards for taking risk-free risks with other people's money – and in the process putting the entire global economy at risk. Meanwhile, other spheres of activity are being suffocated by an excess of regulation. The most egregious example in Britain at the time of writing is the Independent Safeguarding Authority. This new bureaucracy was created as a response to a sensationalist media outcry over the murder of two young girls (the Soham murders in 2002). It is now charged with vetting an estimated 9 million people before they will be permitted to work or volunteer with children or 'vulnerable' adults. The vetting involves a Criminal Records Bureau (CRB) check on all 9 million after which, according to the Home Office website, case workers will "decide on a case-by-case basis whether a person poses a potential on-going risk and if necessary, bar an individual from working with vulnerable groups" (Home Office 2012). Many people must be vetted more than once since the checks relate not just to individuals but to the particular situations in which they might encounter 'vulnerable' people. Between 2002 and 2012, 32 million CRB checks were reportedly conducted at a cost of some £1.5 billion (Beckford 2012).

Leaving aside the mind-boggling expense and bureaucracy required to perform this feat, its effect is almost certain to be perverse. The bureaucratization of the protection of children shifts responsibility. A Criminal Records Bureau check will be seen as an insurance policy; behaviour that might previously have aroused suspicion is now less likely to be questioned, or acted upon, because some superior authority has certified the suspect as 'safe'. But much worse is the damage that will be done by this extraordinarily disproportionate reaction to an extremely rare event. It is already having an impact on volunteering, in a wide range of activities requiring adult involvement. From music and drama to sports, scouting, field trips and educational exchanges, reports suggest a massive withdrawal of adults unwilling to subject themselves to the cost, inconvenience or indignity of the vetting process (Paton 2009).

But still worse, resilience is a skill acquired through experience. Over recent decades in the United States, Britain and many other wealthy countries, the pursuit of zero risk to children has led to their increasing confinement under adult supervision. Now the loss of adult supervisors is restricting still further the range of activities in which they can engage, leaving them to grow increasingly obese in front of their TVs and PlayStations. Learning through experience, the balancing act that underpins resilience, is increasingly denied them.

Resilience has a social dimension. In the aftermath of natural disasters, when emergency services and the forces of law and order are overwhelmed, how people behave will depend on their expectations of the behaviour of others. Members of societies that enjoy a high level of mutual trust are likely to respond cooperatively and altruistically. A society so paranoid that it treats every adult as a potential paedophile until proven otherwise is more likely to respond defensively and selfishly.

Hurricane Katrina provided examples of both types of response. Most ‘ordinary people’ responded selflessly and in many cases heroically. Reports of raping and looting were subsequently revealed to have been grossly exaggerated. But enormous suffering was caused by the paranoia of the official custodians of law and order who impeded, sometimes at gunpoint, informal rescue and evacuation efforts because their default assumption was that the anonymous victims could not be trusted (Solnit 2009).

3.4 Perceptual Filters

The Risk Thermostat of Fig. 3.2 comes equipped with perceptual filters. Cultures and individuals vary widely in their perceptions of the dangers and rewards encountered in the pursuit of resilience. Figure 3.3 proffers a cartoon version of a typology of commonly encountered responses to risk developed in a branch of anthropology called Cultural Theory (Adams 1995).

The *Hierarchist* represents the institutional risk manager, the maker and enforcer of the rules to which society is expected to conform. The ultra-cautious *Egalitarian* in the guise of defender of the environment, or its vulnerable inhabitants, commonly argues that the hierarchy is not doing enough to protect us, whereas the *Individualist*

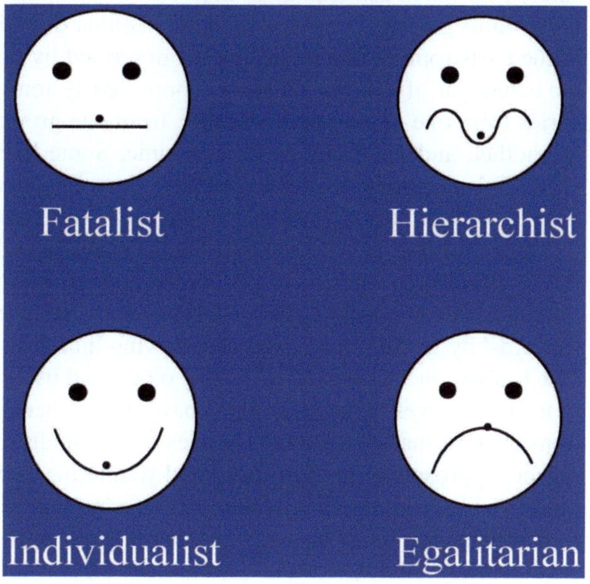

Fig. 3.3 A typology of perceptual filters

complains that the hierarchy is over-regulating and suffocating enterprise and individual liberty. Most of us most of the time are *Fatalists*, conscious of the minimal control that we have over the physical and economic threats that hover over us. We buy lottery tickets and duck if we see something about to hit us.

So far I have proposed that the manner in which individuals and societies respond to risk will depend on what sort of risk they are confronting (Fig. 3.1) and their propensity to take risks (Fig. 3.2) which will depend in turn on their perceptual filters (Fig. 3.3). There is a further complication: the way individuals and societies relate to each other.

3.5 Risk Is an Interactive Phenomenon

We all have a risk thermostat; we all monitor our environment for signs of safety and danger, and adjust our behaviour in response to what we observe. Figure 3.2 can serve as a description of the behaviour of the driver of a single car going around a bend on an empty road. His speed will be influenced by his perception of the rewards of risk: these might range from getting to the church on time, to impressing his friends with his skill or courage. His speed will also be influenced by his perception of the danger: his fears might range from death, through the cost of repairs and loss of his license, to mere embarrassment. His speed will also depend on his judgment about the road conditions – is there ice or oil on the road? How sharp is the bend and how high the camber? It will also be contingent on his perception of the capability of his car – how good are the brakes, suspension, steering, and tires?

Overestimating the capability of the car or the speed at which the bend can be safely negotiated can lead to an accident. Underestimating those things will reduce the rewards gained. The consequences, in either direction, can range from the trivial to the catastrophic. The balancing act described by this illustration is analogous to the behaviour of a thermostatically controlled system. The setting of the thermostat varies from one individual to another, from one group to another, from one culture to another, and for all of these, over time. Some like it hot – a Hell's Angel or a Grand Prix racing driver, for example – others like it cool – a Caspar Milquetoast or a little old lady named Prudence. But no one wants absolute zero.

Figure 3.4 introduces a second (smaller) vehicle to the road to make the point that risk is usually an interactive phenomenon. One person's balancing behaviour has consequences for others. On the road one road user can impinge on another's 'rewards' by getting in his way and slowing him down, or help him by giving way. One is also concerned to avoid hitting other road users or being hit by them. Driving in traffic involves monitoring the behaviour of others, speculating about their intentions, and estimating the consequences of a misjudgement. Drivers who see another vehicle approaching at high speed and wandering from one side of the road to the other are likely to take evasive action, unless perhaps they place a very high value on their dignity and rights as road users and fear a loss of esteem if they are seen giving way. During this interaction enormous amounts of information are processed.

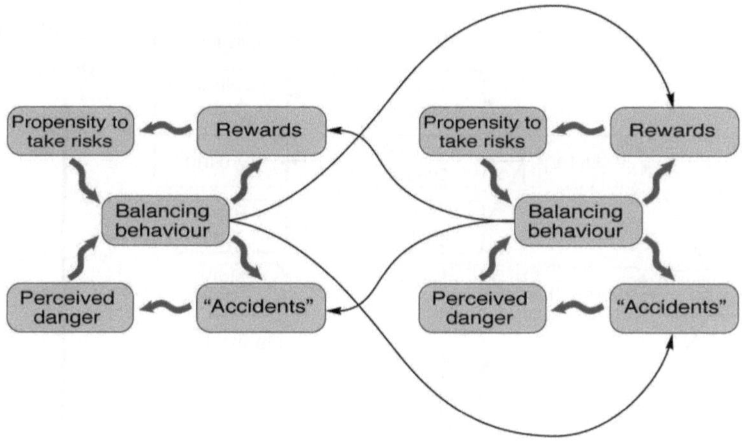

Fig. 3.4 Risk as an interactive phenomenon: the truck driver and the cyclist

Moment by moment, each motorist acts upon information received, thereby creating a new situation to which the other responds.

Figure 3.4 also illustrates a further complication. On the road and in life generally, risky interactions frequently take place on terms of gross inequality. The damage that a heavy truck can inflict on a small car, cyclist or pedestrian is great; the physical damage that a cyclist or pedestrian might inflict on the truck is small. The truck driver in this illustration can represent the controllers of large risks of all sorts. Institutional risk managers who determine the safety of consumer goods, working conditions, or large construction projects are, like the truck driver, usually personally well insulated from the consequences of their decisions. The consumers, workers, or users of their constructions, like the cyclist, are in a position to suffer great harm, but usually not inflict it.

3.6 What Kills You Matters[4]

Whether risks are seen as voluntary or imposed has an enormous influence on the way people respond to them. The terrorist bombs in London on the 7th of July 2005 killed 52 people – the equivalent of 6 days of death on the roads of Britain – but 10,000 people do not gather every weekend in Trafalgar Square to manifest their outrage at the previous week's road death toll with a 3-min silence.

What kills you matters. Figure 3.5 represents an attempt by the author to put in rank order different causes of death according to their acceptability or amplification.

[4] An extended version of this section can be found in Adams (2005).

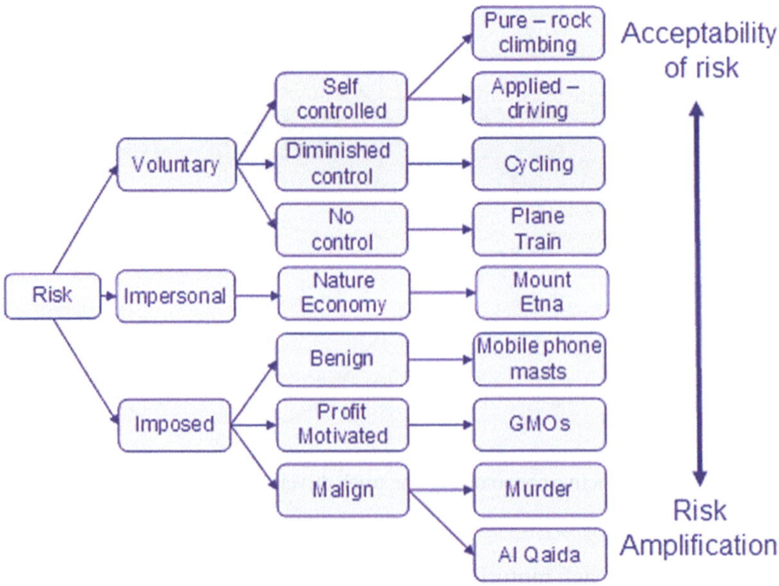

Fig. 3.5 What kills you matters

With 'pure' voluntary risks, the risk itself, with its associated challenge and rush of adrenaline, is the reward. Most climbers on Mount Everest know that it is dangerous and willingly take the risk.

With a voluntary, self-controlled, 'applied' risk, such as driving, the reward is getting expeditiously from A to B. But the sense of control that drivers have over their fates appears to encourage a high level of tolerance of the risks involved.

Cycling from A to B (I write as a London cyclist) is done with a diminished sense of control over one's fate. This sense is supported by statistics that show that per kilometre travelled a cyclist is 14 times more likely to die than someone in a car. This is a good example of the importance of distinguishing between relative and absolute risk. Although 14 times greater, the absolute risk of cycling is still small – one fatality in 25 million kilometres cycled; not even a Tour de France champion can begin to cover that distance in a lifetime of cycling. Moreover, numerous studies (e.g. de Hartog et al. 2010) have demonstrated that the extra relative risk is more than offset by the health benefits of regular cycling; regular cyclists live longer. In some circumstances this risk can be relocated to the 'imposed' part of this diagram. Campaigns for safer cycling are usually based on the perceived injustice of the risks imposed on cyclists by heedless motorists and highway authorities heedless of their 'rights'. Classification in Fig. 3.5 depends on the perspective of the classifier.

While people may voluntarily board planes, buses and trains, the popular reaction to crashes in which passengers are passive victims, suggests that the public demand a higher standard of safety in circumstances in which people voluntarily hand over control of their safety to pilots, or to bus or train drivers.

Imposed risks are less tolerated. Consider mobile phones. The risk associated with handsets is either non-existent or very small (Burgess 2004). The risk associated with the base stations, measured by radiation dose, is orders of magnitude less, unless one is up the mast with an ear to the transmitter. Yet all round the world billions of people are queuing up to take the voluntary risk, and almost all the opposition is focussed on the base stations, which are seen by objectors as impositions. Because the radiation dose received from the handset increases with distance from the base station, to the extent that campaigns against the base stations are successful, they will increase the distance from the base station to the average handset, and thus the radiation dose. The base station risk, if it exists, might be labelled a benignly imposed risk; no one supposes that the phone company wishes to harm all those in the neighbourhood.

Less tolerated are risks whose imposers are perceived as motivated by profit or greed. In Europe, big biotech companies such as Monsanto are routinely denounced by environmentalist opponents for being more concerned with profits than the welfare of the environment or the consumers of its products.

Less tolerated still are malignly imposed risks – crimes ranging from mugging to rape and murder. In most countries in the world the number of deaths on the road far exceeds the numbers of murders, but far more people are sent to jail for murder than for causing death by dangerous driving. In the United States in 2002, the year after 9/11, 16,000 people were murdered – a statistic that evoked far more popular concern than the 42,000 killed on the road but far less concern than for the 25 killed that year by terrorists.

This brings us to terrorism and al-Qaeda. How do we account for the massive scale, worldwide, of the outpourings of grief and anger attached to its victims, whose numbers are dwarfed by those of other causes of violent death? Up to this point we have been discussing individual responses to a range of risks. Terrorism targets *governments*. Terrorists pose a threat not just to individuals but to the social order – *and* to those who purport to maintain it. Murderers and careless drivers are not seen as threats to the ability of the government (the Hierarchy) to govern. And governments have multitudes of press officers and IT experts to amplify their anxieties.

In the middle of this scale are 'impersonal risks', risks imposed by nature or impersonal economic forces. They are commonly considered to be neither voluntary nor imposed. High impact, low frequency risks such as the 2004 Indian Ocean Tsunami or asteroid impact are seen as beyond the control of individuals or governments and responded to fatalistically.[5]

But often the impersonal risks become personal. Mount Etna was chosen to represent such risks in Fig. 3.5 because the diagram was originally prepared for a conference in Sicily. However, I was politely informed by Sicilians at the conference that Mount Etna was a very friendly volcano. Over thousands of years, I was told, only 77 deaths had been attributed to its activity, and most of these could be assigned to the voluntary

---

[5] Although reports appear from time to time of rocket scientists proposing that with sufficient funding they could devise ways of deflecting them (e.g. Matson 2009).

risk category: either viticulturists and farmers exploiting the rich volcanic soils, or curious tourists and geologists venturing too close to the edge.

On closer inspection many 'impersonal' economic risks, while beyond the control of individuals or governments, also turn out to be not impersonal at all. The sub-prime crunch which began with the bursting of house price bubble in 2007 triggered a worldwide recession and devastated the economies of smaller countries such as Iceland, Ireland and Greece. The response has ranged from impotent rage against 'the bonus culture' to, in a few cases, criminal prosecutions.

3.7 The Dance of the Risk Thermostats

Figure 3.6 is an attempt to bring together all the complications discussed above. The world contains more than 7 billion (and growing) risk thermostats. Some are large – presidents with fingers on buttons; most are tiny – children chasing balls across streets or shepherds in Afghanistan. Governments and big businesses make decisions that affect millions if not billions of people. Individuals for the most part adapt as best they can to the consequences of those decisions. The damage that they individually can inflict in return, through the ballot box or market, is insignificant, although in aggregate they can become forces to reckon with.

The broken line symbolizes the uncertain impact of human behaviour on nature. Overhanging everything are the forces of nature – floods, earthquakes, hurricanes, plagues – which even governments cannot control, although they sometimes try to build defences against them. And fluttering about the dance floor are the Beijing butterflies beloved of chaos theorists: they ensure that the best laid plans of mice and men 'gang aft agley'.

The small winged creature at the top left of Fig. 3.6 was added in response to a survey in *Time* magazine (23 December 1993) that revealed that 69% of Americans believed in angels and 46% believed they had their own guardian angel. The 'angel factor' must influence many risk-taking decisions – in mysterious ways. At the time that I added the angel I thought she was amusing. Subsequently, pondering the role of religious filters, God and his angels appear to be assuming growing salience. In the War on Terror, God appears to intercede on both sides with devastating effect.

Amongst the 7 billion thermostats, all the different perceptual filters discussed above in Fig. 3.3 will be represented, all the categories of relationship described in Fig. 3.4 will be found, and they will all be forced to confront the different types of risk presented in Fig. 3.1. Figure 3.6 shows but a minute fraction of the possible interactions between all the world's risk thermostats, and the complexity of these relationships is increasing rapidly. As the world's population increases and grows more mobile, both physically and electronically, the number and length of the connections linking thermostats increases.

Can such a system be 'managed'? In the worlds of commerce and industry there is a spreading fashion amongst large companies, especially in the financial sector, to appoint CROs – Chief Risk Officers. This new office appears to have been created in

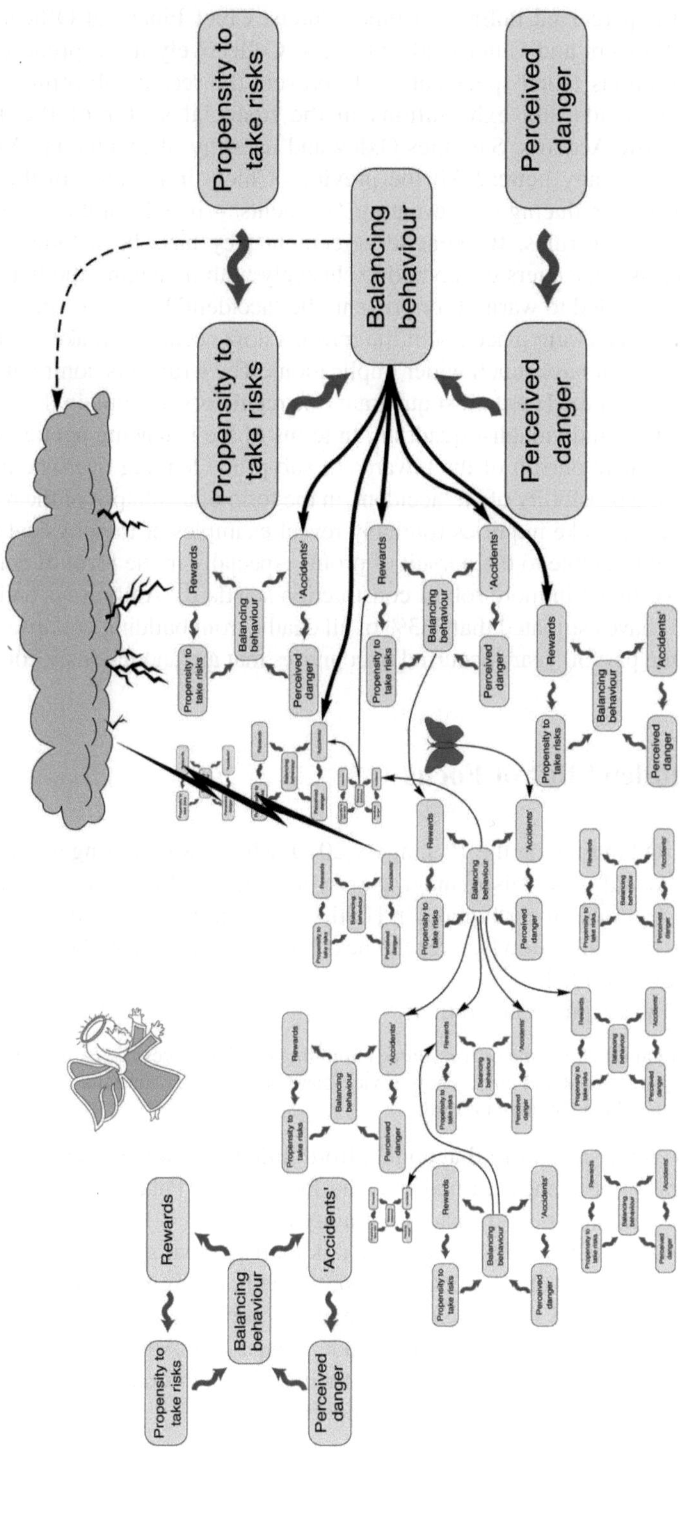

Fig. 3.6 The dance of the risk thermostats

response to the perceived failings of other Chiefs: Chief Financial Officers, Chief Compliance Officers and Chief Audit Officers. Collectively these predecessors in financial institutions failed spectacularly to prevent the recent sub-prime crunch – despite the demands and exhortations in the financial sector of the Turnbull Report, the Basle Accords, Sarbanes Oxley and an army of regulators. Will Chief Risk Officers fare any better? All the previous Chiefs implicated in the Crunch were charged with reducing or preventing 'accidents' – mostly in the form of non-compliance with the rules. By general agreement they have been failed 'bottom-loopers' – i.e. risk managers concerned exclusively with managing the bottom loop of Fig. 3.2; they failed to warn of, or prevent, the 'accident' known as the Crunch.

The discussions taking place about the reforms now needed to make the financial sector more resilient have much wider implications. The Crunch is commonly attributed to the failure of the Hierarchist quadrant (the regulators) to contain the 'irrational exuberance' of the Individualist quadrant. In terms of the balancing act described by Fig. 3.2, the frenetic pursuit of the rewards of sub-prime lending overwhelmed contemplation of the possibility of an 'accident' in the form of a collapse of the real estate market. Post-earthquake inquiries routinely reveal examples of the physical collapse of real estate attributable to the pursuit of profit, especially in the form of corruption, overriding investment in more robust construction standards. Ambraseys and Bilham (2011, p. 153) have estimated that "83% of all deaths from building collapse in earthquakes over the past 30 years occurred in countries that are anomalously corrupt".

3.8 A Sudden Shift of Focus

At 11 p.m. London time on the 12 January 2010, while I was fretting about the way that the rich world was (mis)managing risk, my radio, which was playing in the background, reported an earthquake in Haiti – a poor faraway country about which I knew almost nothing. Slowly, over the next few hours and days, the magnitude of the devastation unfolded.

Earlier in the chapter, I noted that:

> Resilience requires command over resources. Building flood defences and earthquake resistant buildings, accident and emergency services, and post-disaster continuity planning are all luxuries that the poor cannot afford.

Haiti was clearly a country that could afford little in the way of earthquake resistant buildings – not even, it turned out, the Presidential Palace – or accident and emergency services, or post-disaster continuity provision.

Can a connection be made between the risk management systems put in place in rich countries to protect children from the statistically negligible risk posed by murderous paedophiles and the systems that exist in poor countries to protect people from risks that are many orders of magnitude greater? Do Amsterdam and Port-au-Prince have anything in common? Despite their different circumstances, do rich and poor countries have common lessons to impart?

All of the inhabitants of Amsterdam and Port-au-Prince – and the rest of the world – are participants in the dance illustrated by Fig. 3.6. All the biases depicted in Fig. 3.3 can be found on the dance floor. Individualists are dancing to tunes of their own choice conveyed through earphones from their iPods; they display a preference for raucous authority-challenging punk rock. The more communally inspired Egalitarians might be engaged in square dancing or line dancing. The Hierarchy is suspicious of dancing and bans it completely in some theocratic cultures; where permitted they display a preference for music appropriate to marching rather than dancing. The poor Fatalists can be recognized by their slow despondent shuffle.

Over different parts of the global dance floor the proportions of the participants exhibiting these biases vary greatly. I insisted above that there is no agreed numerical measure of resilience. There are, however, some helpful pointers. On a rank-order resilience scale the Netherlands would be close to the top and Haiti close to the bottom. Life expectancy at birth in Haiti is 62.5 years and in the Netherlands 80.9 years (CIA 2012a). On a daily basis some combination of accident, disease, malnutrition, accident and occasional disaster is depriving the average Haitian of over 18 years of life. Judged by this indicator, Haiti must be less resilient.

In the discussion above of Figs. 3.4 and 3.6 it was observed that the interactions between individual risk managers often take place on terms of gross disparity. The Haiti tourism industry provides an extreme example. A few days after the earthquake, a Royal Caribbean cruise liner arrived at the private resort of Labadee, 85 miles from Port-au-Prince. There was much discussion in the press about the unseemly nature of rich tourists disporting themselves in such close proximity to tragedy (Sequera 2010). But a spokesman for the travel agency justified the visit: "Anything you can do to continue normalcy in Haiti is helpful… People will feel they're doing more by going there than not." (Theme Park Review 2010). What "normalcy" means in Labadee when tourist meets Haitian is nicely captured by a tourist's travel diary posted on the Internet before the earthquake:

> There is a small section for kids with floating fake icebergs to climb and water slides but they charge an admission fee. Flotation mattresses are also available for rent for those who just want to float in the ocean. There are also locals who help with getting beach chairs for you but they expect tips. So things here are a money grabber (…) One big difference between this port compared to others is that since it is pretty well being used as a private beach, you will not be hassled by locals. The vendors coming around with drinks at the beach locations are actually Royal Caribbean staff so if one wishes to buy a drink, the cruiseship passcard is all that is required. As for concerns about being in Haiti given the poverty and political situation, Labadee was not a problem because the entire site is enclosed by a high steel fence. Other passengers later told us that they wandered off near the perimeter of the site and saw many Haitians along the fence begging for handouts and food. The fencing was mostly concealed in the distance from the main tourist areas (CruiseJournals 2005, comment posted 2005).

There exist two further international scales by which Haiti and the Netherlands might be compared that have a bearing on resilience. One is corruption. A resilient society will be cooperative; people will be prepared to help each other. In corrupt societies assistance is for sale, not freely given. For most of its history the *hierarchy* ruling Haiti has been a corrupt tyranny. In Transparency International's (2010) survey of corruption the Netherlands ranked 7th least corrupt nation in the world. Haiti tied with seven other countries at 146.

A second relevant indicator is economic equality. Billionaire bankers and shop floor workers or 'slum dogs' do not have many reasons to pull together in the face of adversity. Various international comparisons of income distribution produce different rankings, but most of north-western Europe, despite bankers' bonuses, can be found at the equitable end of the international equality scales while Haiti sits at the bottom, along with a group of African kleptocracies (CIA 2012b).

In post-earthquake Haiti, religious belief (represented by the angel in Fig. 3.6) appears to have played a significant role in helping many Haitians, condemned to fatalism by their poverty, to cope with the devastation in which they found themselves. Television and other reports of prayer meetings, the singing of hymns, and voodoo observances suggest that various forms of faith can be the basis of a form of resilience. How might a population of impoverished atheists cope with the devastation inflicted on Haiti? Do such populations exist? Or does the struggle for continued existence in the face of disaster and desperate poverty support, or induce, belief in the supernatural? Such questions are beyond the competence of the author of this chapter, but they do serve to illustrate the dense complexity confronting an investigation of resilience.

Figure 3.2 was proffered above as an example of the risk-decision making process of a driver going around a bend in the road. It was deployed as an example from a rich country in which most people ride around in cars. Can it be applied to Haiti? Everyone in Haiti, as I write, is a risk manager. Whether they have just crawled out from the ruins of a home, office, hotel, factory, school or shop, whether they are a policeman, government minister or aid worker, they are all peering into a desperately uncertain future wondering what to do next. The decision about the speed at which to negotiate a bend in the road is exchanged in Haiti for a decision about whether to scrabble through the ruins of a local shop to acquire life-sustaining food and drink at the risk of being shot as a looter.

The driver or 'looter' in deciding what to do is compelled to anticipate a reaction to their behaviour. The population of Haiti is about the same size as the population of that part of the Netherlands susceptible to flooding. Both populations are vulnerable to natural and human-made hazards. Every individual inhabitant is a risk manager with a risk thermostat that has a perceptual filter.

Who should be in charge of the *societal* balancing act? Moreover, how is the societal risk thermostat set, and what sort of risk should the thermostat setter contemplate?

3.8.1 Who Is in Charge?

In rich or middle-income countries in the case of flood or earthquake risks, it is those branches of government responsible for flood defences, building codes and emergency services. Insurers also have a role to play in the financing of reconstruction. In Haiti, there were no building codes relevant to earthquake risk and no emergency services that could make a perceptible impression on the disaster. Both the presidential palace and any semblance of central control collapsed. Initial efforts by international aid agencies and the 9,000 uniformed UN personnel present in the country to impose some sort of order were seriously undermined by large numbers of escaped prisoners, and the lack of any effective policing.

In 1989, the San Francisco Bay area was struck by a magnitude 7 earthquake, the same magnitude as the quake that killed 250,000 people[6] and devastated the economy of Haiti. It killed only 63 and left barely an economic ripple (Eberhart-Phillips et al. 1994). The *insured* losses were much greater in California because few in Haiti could afford insurance and the monetary value of the properties destroyed was negligible in the eyes of international insurance companies. In Chile, on 27 February 2010, a much stronger earthquake accompanied by a tsunami killed only an estimated 500, and the president was promptly on the streets and on television clearly in command of rescue operations.

3.8.2 How Is the Thermostat Set?

With great difficulty. In addition to the problems of poverty and corruption discussed above there is the unresolved problem of deciding the magnitude of the resources that should be committed to preparing for extremely low frequency, high impact events. In California and Chile, earthquakes are not only high impact events, they are also frequent events. In Haiti the last earthquake of similar magnitude is reported to have occurred 250 years ago (Bajak 2010). For a desperately poor country, planning for such a rare event is understandably not a high priority.

3.9 Whose Appetite for Risk?

Commenting on the aftermath of the Haiti earthquake in *The Telegraph* (2010), Elisabeth Byrs, spokeswoman of the UN Office for the Coordination of Humanitarian Affairs, stated, "This is a historic disaster. We have never been confronted with such a disaster in the UN memory. It is like no other". However, as noted above, poverty kills, and those hit hardest by the recession triggered by the sub-prime credit crunch of 2007 are the poorest. The reckless risk-taking of so-called financial risk managers may yet be credited with causing more deaths than those caused directly by the Haiti earthquake.

Financial risk managers like to trade slogans such as 'no risk, no reward' and 'no pain, no gain' and to advise each other that they should 'manage risk within their

---

[6] This is a widely reported estimate. The true number will never be known because large numbers were buried in mass graves with no one keeping count.

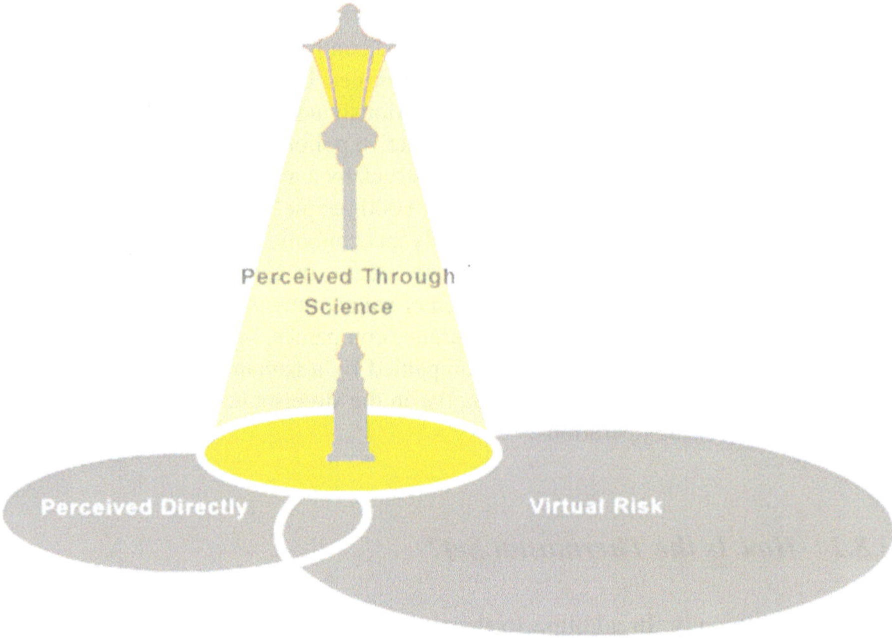

Fig. 3.7 The pursuit of resilience. Where are the keys?

organisation's risk appetite'. They are right. A risk-free world is not possible and, as we have seen above, excessive risk aversion incurs costs in the form of potential rewards foregone. The key to a fair system of societal risk management is to ensure that those who collect the rewards share the pain when things go wrong. Some financial risk managers, still collecting their bonuses, appear, like the tourist on the cruise liner introduced above, to be well insulated from the pain suffered by those less fortunate.

3.10 Where Are the Keys?

A final note of caution. The mythical drunk notoriously searches for his keys not in the dark where he dropped them, but under the lamppost where he can see (Fig. 3.7). This is an apt metaphor for much of what is written on the subjects of resilience and risk management.

Lord Kelvin famously said, "Anything that exists, exists in some quantity and can therefore be measured" (as quoted in Beer 1967, p. 14). This dictum sits challengingly alongside that of another famous scientist, Peter Medawar (1967, p. 11) who observed, "If politics is the art of the possible, research is the art of the soluble. Both are immensely practical minded affairs. Good scientists study the most important *problems they think they can solve* [my italics]. It is, after all, their professional business to solve problems, not merely to grapple with them".

Risk is a word that refers to the future. The future has no objective existence. It exists only in the imagination. There are some risks for which science can provide useful guidance to the imagination. The risk that the sun will not rise tomorrow can (I hope) be assigned a very low probability by science. Actuarial science can estimate with a high degree of confidence that the number of people killed in road accidents in Britain next year will be 2,000, plus or minus a hundred or so. But these are predictions, not facts. Such predictions rest on assumptions: that tomorrow will be like yesterday; that next year will be like last year; that future events can be foretold by reading the runes of the past. Sadly, the history of prediction contains many failures – from those of stock market tipsters to those of scientists seeking to predict eruptions, earthquakes and tsunamis.

In the area lit by the lamp of science, one finds risk management problems that are potentially soluble by science. Such problems are capable of clear definition relating cause to effect and characterised by identifiable statistical regularities. On the margins of this circle one finds problems framed as hypotheses, and methods of reasoning, such as Bayesian statistics, which guide the collection and analysis of further evidence. As the light grows dimmer, the ratio of speculation to evidence increases. In the outer darkness lurk unknown unknowns. Here lie problems with which, to use Medawar's word, we are destined to 'grapple'.

References

Adams J (1974) …and how much for your grandmother? Envir Plan A 6(6):619–626

Adams J (1985) Risk and freedom: the record of road safety regulation. Transport Publishing Projects, London

Adams J (1995) Risk. UCL Press, London

Adams J (2005) What kills you matters, not numbers, the social affairs unit. http://www.socialaffairsunit.org.uk/blog/archives/000512.php. Accessed 7 Aug 2012

Ambraseys N, Bilham R (2011) Corruption kills. Nature 469:153–155

Bajak F (2010) Chile-Haiti earthquake comparison: Chile was more prepared. http://www.huffingtonpost.com/2010/02/27/chile-haiti-earthquake-co_n_479705.html. Accessed 29 Mar 2011

Beckford M (2012) CRB checks top 30 million but create 'atmosphere of mistrust'. The Telegraph, 23 Apr 2012. http://www.telegraph.co.uk/news/uknews/law-and-order/9219430/CRB-checks-top-30million-but-create-atmosphere-of-mistrust.html. Accessed 9 Aug 2012

Beer S (1967) Management science. Aldus, London

Borthwick A (2011) Coastal flood risk – from storm surge and waves to inundation. http://web.sbe.hw.ac.uk/frmrc/downloads/Coastal_Flood_Risk_London/Alistair%20Borthwick.pdf. Accessed 7 Aug 2012

Burgess A (2004) Cellular phones, public fears, and a culture of precaution. Cambridge University Press, Cambridge

CIA (2012a) Country comparison: life expectancy at birth. https://www.cia.gov/library/publications/the-world-factbook/rankorder/2102rank.html. Accessed 7 Aug 2012

CIA (2012b) Distribution of family income: GINI index. https://www.cia.gov/library/publications/the-world-factbook/rankorder/2172rank.html. Accessed 7 Aug 2012

CruiseJournals (2005) A day in Labadee at Hispaniola, Haiti. http://www.cruisejournals.com/read-article.php?article_id=22. Accessed 1 June 2012

de Hartog JJ, Boogaard H, Nijland H, Hoek G (2010) Do the health benefits of cycling outweigh the risks? Environ Health Perspect 118:1109–1116

Deltawerken Online (2011) The delta works. http://www.deltawerken.com/23. Accessed 29 Mar 2011
Eberhart-Phillips JE, Saunders TM, Robinson AL, Hatch DL, Parrish RG (1994) Profile of mortality from the 1989 Loma Prieta earthquake using coroner and medical examiner reports. Disasters 18:160–170
Home Office (2012) The independent safeguarding authority. http://www.homeoffice.gov.uk/agencies-public-bodies/crb/partners-reg-bodies/isa/. Accessed 9 Aug 2012
Jonkman SN, Stive MJS, Vrijling JK (2005) New Orleans is a lesson to the Dutch. J Coast Res 21(6):11–12
Matson J (2009) Russian space chief makes vague threat to vaguely threatening asteroid. http://www.scientificamerican.com/blog/post.cfm?id=russian-space-chief-makes-vague-thr-2009-12-30. Retrieved 29 Mar 2011
Medawar PB (1967) The art of the soluble: creativity and originality in science. Methuen, London
Moss S (2011) Weatherwatch: the grote mandrenke. http://www.guardian.co.uk/news/2011/jan/20/weatherwatch-grote-mandrenke. Accessed 7 Aug 2012
Paton G (2009) CRB checks 'deterring school volunteers.' http://www.telegraph.co.uk/education/6281842/CRB-checks-deterring-school-volunteers.html. Accessed 7 Aug 2012
Schenau S, Delahaye R, Edens B, van Geloof I, Graveland C, van Rossum M, Wolswinkel KJ (2009) Milieurekeningen 2008. Centraal Bureau voor de Statistiek. http://www.cbs.nl/nl-NL/menu/themas/natuur-milieu/publicaties/publicaties/archief/2009/2008-c167-pub.htm. Accessed 7 Aug 2012
Sequera V (2010) Haiti cruise stops: 'without this, we don't eat'. The Guardian, 23 Jan 2010. http://www.guardian.co.uk/world/feedarticle/8912655. Accessed 9 Aug 2012
Solnit R (2009) A paradise built in hell: the extraordinary communities that arise in disaster. Viking, New York
The Telegraph (2010) Haiti earthquake: police open fire on looters. http://www.telegraph.co.uk/news/worldnews/centralamericaandthecaribbean/haiti/7005853/Haiti-earthquake-police-open-fire-on-looters.html. Accessed 7 Aug 2010
Theme Park Review (2010) http://www.themeparkreview.com/forum/viewtopic.php?f=2&t=54088&start=30. Accessed 5 Feb 2010
Transparency International (2007) Corruption perceptions index. http://www.transparency.org/policy_research/surveys_indices/cpi/2007. Accessed 29 Mar 2012
Transparency International (2010) Corruption perceptions index 2010 results. http://archive.transparency.org/policy_research/surveys_indices/cpi/2010/results. Accessed 7 Aug 2012
van Alphen J (2009) Flood risk management in the Netherlands. Organisation, legislation and governance. http://www.nfrmp.us/ifrma/docs/pre/governance/NetherlandsGovernancePaper.pdf. Accessed 7 Aug 2012
Wikipedia (2012) Delta works. http://en.wikipedia.org/wiki/Delta_Works. Accessed 9 Aug 2012
Wilde G (1976) The risk compensation hypothesis theory of accident causation and its practical consequences for accident prevention. Paper presented at annual meeting of Österreichische Gesellschaft für Unfallchirurgie, Salzburg, 1976

Chapter 4
Responding to Flood Risk in the UK

Tim Harries

Abstract This chapter considers the response of UK householders to the country's most widespread and damaging natural hazard, flooding. Although flood risk affects 3 million UK residents and major floods in 1998, 2000, 2005, 2007 and 2009 received extensive media coverage, few at-risk householders take any action to reduce their risk exposure. Research conducted in London, Reading and Leeds suggests that people who have insufficient confidence in their ability to manage their exposure to the material impacts of flooding choose instead to adopt anxiety-avoidance strategies such as blame and fatalism. These strategies protect social representations that enable citizens to achieve a feeling of safety in their lives but they also de-legitimise the discourse of risk mitigation. The research suggests that protection of self-identity and social identity also play a role. Only when traumatic or repeated experiences of flooding force changes to identity and make the retention of old representations untenable are these psychological strategies abandoned. When this occurs, individuals either learn to accept the existence of the risk or else fall into a state of disabling anxiety.

4.1 Introduction

Although 60% of at-risk residents in England and Wales claim to be aware that they live in flood risk areas, only 39% of those with experience of flooding and 6% with no such experience have taken any action to reduce their risk exposure (Harries 2008a). This chapter asks why this is so. It argues that people's apparently irrational refusals to prepare for flooding are functional when seen in the context of a lack of

T. Harries (✉)
Behaviour and Practice Research Group, Business School,
Kingston University London, Kingston-upon-Thames KT2 7LB, UK
e-mail: t.harries@kingston.ac.uk

H. Joffe et al. (eds.), *Cities at Risk: Living with Perils in the 21st Century*, Advances in Natural and Technological Hazards Research 33,
DOI 10.1007/978-94-007-6184-1_4,

confidence in the available mitigation measures and their assessments of the negative effects associated with them. It argues that people fail to protect themselves when they doubt their ability to do so effectively and fear that adaptive action will have a detrimental impact on anxiety levels and will threaten self-esteem and the sense of belonging.

Flooding is the UK's most common and costly form of natural disaster. Over 400,000 households and 1.5 million people in England and Wales have a greater than 1.3% annual chance of having their homes flooded by a fluvial or tidal flood event (Defra 2008a; Evans et al. 2004) and perhaps as many again are at risk from flooding caused by the incapacity of drainage systems to cope with heavy rain.

After decades of relatively few floods of national significance, in the last fifteen years numerous events have attracted the interest of national media and policy-makers. In 1998, the 'flood of the century' (Guardian 1998) affected thousands of homes and businesses across England and Wales, causing £300 million of damage. Two years later, heavy rain in already saturated areas led to flood damage of £1bn and the evacuation of 11,000 people across England and Wales (Johnson et al. 2008). In 2005 and 2009, large parts of the county of Cumbria were flooded. Finally, in 2007, the first nationally significant event caused by surface water flooding led to a second 'flood of the century' (Observer 2007) that affected 48,000 homes and caused £6 billion of damage across English towns and cities.

These events contributed to an emerging consensus amongst flood professionals that the frequency of flooding and the extent of exposure were increasing faster than existing management strategies could cope with. An investigation into the impacts of climate change (Evans et al. 2004) concluded that if expenditure on flood defence was maintained at existing levels, flood damage in England and Wales would rise to between £2 billion and £30 billion per year by the 2080's and the number of people living in high risk areas would increase from 1.6 million in the year 2000 to between two and four million.[1] Furthermore, it was recognised that demographic change, demand for housing, and policies favouring the development of river catchment areas and brown-field sites were maintaining the pressure for more homes to be built on at-risk land (Evans et al. 2004; McCarthy et al. 2001; Smith and Ward 1998). At the same time, there was growing awareness of the health impacts of flooding, with research suggesting that a third of flood-victims suffer long-term adverse physical effects (Tapsell et al. 1999; World Health Organisation 2002) and that the anxiety, relationship strain and general disruption that comes in the wake of flooding is associated with increases in mental ill-health (Tunstall et al. 2006).

As a result of these developments, the UK government recognised that engineered flood defences alone were inadequate to the task of mitigating the nation's flood risk and that a "portfolio" of approaches was needed (Defra 2005, p. 8). This "portfolio" included steps that householders themselves could take (Johnson et al. 2005) – i.e. measures to slow or prevent the ingress of floodwater into individual

[1] The large differences between estimates are the result of the use of different assumptions about future economic systems and policies and different assumed levels of economic growth.

Fig. 4.1 Examples of flood barriers: (**a**) sandbags; (**b**) a home-made door-board made of marine ply, and (**c**) commercially available airbrick covers (*Source*: The author)

homes (Fig. 4.1) or reduce the damage incurred when water does gain ingress (e.g. the use of water-resistant fixtures and fittings). However, early attempts to promote the use of these physical mitigation measures were far from effective (Pitt 2008) and by 2008 only approximately 10% of households in high risk areas had implemented any such steps (Harries 2008a).

4.2 The Research

Research on the barriers and incentives for the implementation of property-level mitigation measures was conducted by the author between 2004 and 2010 across a range of urban and rural areas of England and Wales, using both qualitative and

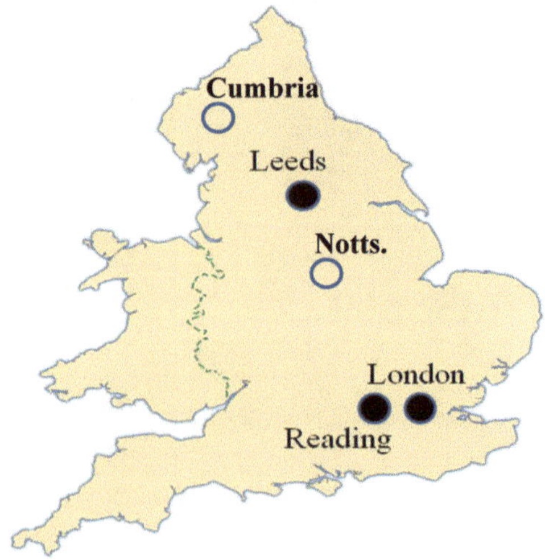

Fig. 4.2 Map of England and Wales showing the fieldwork locations for the qualitative data collection (*Source*: The author)

quantitative methodologies. This work was sponsored by the Economic and Social Research Council, the Environment Agency and the Department for Environment, Food and Rural Affairs (Defra), with additional support from King's College London and the Flood Hazard Research Centre at Middlesex University.

4.2.1 Qualitative Research

The qualitative research analysed the language and arguments that householders use when they talk about flooding and flood risk. It aimed to identify not only expressed justifications for taking or not taking mitigating measures, but also underlying reasons for these justifications and what they revealed about people's hopes, fears and fundamental motivations.

To this end, exploratory 'semi-structured' interviews and focus groups were conducted with at-risk and flooded householders in six parts of England, including the cities of London, Reading and Leeds (Fig. 4.2).

Many parts of London's ethnically diverse population of 7.5 million are threatened by flooding. In the centre of the UK's capital, both the financial City and the political quarter lie within the floodplain of the Thames. However, central London was last flooded in 1928 and now benefits from a high standard of flood defence, so most fluvial flooding occurs in the suburbs and is largely caused by the overtopping of urban streams, many of them canalised and some of them also acting as sewers (Fig. 4.6). For example, in 2000, the overtopping of defences on the River Roding in north-east London caused flooding to 230 properties (London Assembly 2002). An additional 680,000 properties are at risk of pluvial flooding due to the incapacity of ancient drainage systems in coping with heavy rainfall events (Fig. 4.3) (Greater London Authority 2010).

Fig. 4.3 (**a**) A typical suburban London street. (**b**) A nearby canalised urban stream that is prone to flooding (*Source*: The author)

Flooding from urban watercourses and surface water is hard to predict. Residents often receive little warning but floodwater dissipates quickly, so seepage through brickwork and floors is not usually a problem. The London research reported here was conducted in one area that suffers from run-off flooding after severe rain and one that experiences flooding from a canalised urban stream. Both were last flooded a few years before the research and neither is threatened by other natural or industrial hazards.

With a population of over 230,000 the second urban area, Reading, is one of London's largest commuter satellite towns. Parts of Reading are flooded from the River Kennet and the River Thames. Although wide-spread flooding occurs

Fig. 4.4 Thames-side properties in Reading (*Source*: The author)

relatively infrequently (1894, 1910, and 1947), two smaller events during the past ten years have raised fears that the frequency might be increasing. Qualitative research was conducted in two areas near the Thames: a newly-built estate of social housing set back from the river and an area of private housing abutting the river (Fig. 4.4).

The third urban fieldwork area is in Leeds, a city of 800,000 residents that forms part of the West Yorkshire metropolitan area. A major manufacturing centre in the eighteenth and nineteenth centuries, Leeds experienced economic decline before emerging as a centre for tertiary activities such as call centres and corporate offices. Interviews were conducted in a small estate of semi-detached, mid-value private homes in a city suburb where there had been three floods in the previous 10 years. Residents had actively lobbied to have the nearby dredged and widened and had recently received door-boards and airbrick covers as part of a government pilot scheme.

Householders in all the areas were recruited on the door-step and offered small cash incentives for their participation. In all, 50 residents of urban flood risk areas were interviewed in focus groups; paired interviews and a one-to-one interview. The composition of this sample is shown in Table 4.1.

Interviews were transcribed verbatim and analysed using a technique developed from *textually based discourse analysis* (Fairclough 2003) and *discursive psychology* (Potter and Wetherell 1987). This analytical method sees language as performative as well as communicative and as constituting reality, not just describing it (Austin 1962; Halliday 1994; Wittgenstein 1958). Analysts sensitise themselves to the different rhetorical and linguistic strategies that might be used and read texts with these strategies in mind, uncovering constructions and intentions that might otherwise be overlooked. They critically interrogate their own presuppositions and unexamined techniques of sense-making, asking, "Why am I reading this passage in this way?" and "What features [of the text] produce this reading?" (Potter and Wetherell 1987, p. 168).

Table 4.1 Characteristics of participants in the qualitative research

Social grade[a]	ABC1 – 18	C2DE – 32	–	–	–
Housing tenure	Owner-occupiers	Social tenants	Private tenants	–	–
	36	11	3	–	–
Household composition	Living alone	With friends/relatives	With partner only	With partner and children	Single parent
	12	2	18	15	3
Flood experience	Direct experience	Experienced near-miss event	Home flooded while away	Never flooded	–
	19	22	2	7	–
Implementation of mitigation measures	Own measures to keep water out	Own measures to reduce vulnerability of fixtures/fittings	Grant-funded mitigation	No mitigation	–
	11	8	10	21	–

[a]'ABC1' refers to people engaged in non-manual work, including professionals, managers and owners of small businesses. 'C2DE' refers to manual workers and those with long-term dependency on the state (see Market Research Society 2002)

4.2.2 Quantitative Research

Findings from the early phases of the qualitative research were used to inform the design of a telephone survey commissioned by Defra in 2007 (see Thurston et al. 2008). In this survey (known from now on as *Survey 1*), people living in flood risk areas were asked whether they had used any of a range of mitigation measures and whether they agreed with a selection of statements describing barriers and facilitators to mitigation (Fig. 4.7). The survey sample was drawn from lists of telephone numbers provided by data supply companies for postcode areas with a greater than 80% concentration of properties in areas where the annual probability of flooding was at least 1.3% – the level of risk classified as 'significant' by UK government bodies. As the research was focused on the actions of people already aware of the risk, a question at the start of the survey was used to screen out those not previously aware ("Before we approached you to take part in this survey, did you believe your home to be at risk of flooding?"). The achieved sample of 555 respondents represented a completion rate of 28%. Less than 10% of these 555 reported having taken any kind of property-level measure to reduce their exposure to flood risk.

Quantitative findings are also reported here from secondary analyses of two further Defra surveys. The first, *Survey 2* (collected in 2002), involved 1,400 face-to-face interviews with respondents in flood risk areas, all of whom were aware of the risk and 1,000 of whom had experienced household flooding (see RPA et al. 2004). The second, *Survey 3* (collected in 2005), involved 280 householders, 94% of whom had experienced flooding (see Tunstall et al. 2006). Respondents in all three surveys were asked whether they had implemented any flood risk mitigation measures and those in surveys 1 and 2 were also asked about the barriers and incentives for such actions (Fig. 4.5).

Survey data was analysed using chi-square tests and multivariate logistic regression – the former to narrow the range of possible predictors of adaptive behaviour; the latter to reduce the effect of spurious associations and discriminate between direct associations and associations via intervening variables (Bohrnstedt and Knoke 1984).

The findings of both the qualitative and quantitative elements of this research are outlined below. Section 4.3 describes the importance of the perception of mitigation measures – their cost, the implications of their use on property prices and their reliability – before Sects. 4.4, 4.5, and 4.6 describe some of the less obvious barriers to the use of these measures. Section 4.4 suggests that people who lack confidence in their ability to choose the right mitigation measures sometimes focus, instead, on reducing the feeling of risk. Section 4.5 considers the influence of identity and trust on a second determinant of take-up levels: the attribution of responsibility for the management of flood risk. Finally, Sect. 4.6 argues that the strategies described in Sects. 4.4 and 4.5 become less tenable with increased experience of flooding and that people who experience particularly numerous or traumatic floods either become psychologically resilient to the risk or else fall into a state of debilitating anxiety.

People have given reasons for NOT putting in place measures to minimise the damage to their homes from flooding. I'm going to read out a list of these reasons. Please say whether you agree, disagree or don't know.

... I feel it would be too expensive

... I don't think it's my responsibility

... I don't want to be reminded of the risk of flooding

... If I'm selling my home, I don't want people to see it's at risk of flooding

... My home is covered by insurance so I don't need to worry

Some people prefer to put in place measures to prevent or minimise damage to their homes from flooding. I'm now going to read out a list of reasons that they have given for having this attitude. Please say whether you agree, disagree or don't know.

... It would make me feel safer

... My insurance premiums would go down or not go up so much

... It would increase the value of my property

... It would decrease the hassle/disruption if there was a flood

Fig. 4.5 Survey questions regarding perceptions of flood risk mitigation

4.3 Perceptions of the Mitigation Measures – Cost, Stigma and Reliability

Although this chapter focuses on the latent drivers of mitigation, it is important to recognise the existence of the overt justifications that people give for their behaviours.

For example, the low uptake of mitigation measures is sometimes attributed to their cost. A full set of commercially available protection measures costs between £2,900 (Defra 2008b) and £4,000 per property (Harries 2010a). However, although 57% of respondents in Survey 1 said they were deterred from implementing mitigation measures because they believed they would be "too expensive", there was no statistical association between perceptions of cost ("I feel [mitigation] would be too expensive" – agree/disagree/neither) and the adoption of such measures (χ^2 [2, $N=519$]$=4.23$, $p>.05$). Furthermore, participants in the semi-structured interviews rarely mentioned cost as a factor unless prompted (Harries 2008a).

A second practical issue is concern over effects on real-estate value. The permanent and visible fittings required by some protective barriers (Fig. 4.6) draw attention to the fact that a property is at risk; as do features such as raised electricity sockets. Twenty-four percent of owner-occupiers hesitate to take adaptive measures in order to avoid revealing the flood risk to prospective buyers (Survey 1) and this view is negatively related to the implementation of such measures (χ^2 [2, $N=431$]$=7.17$, $p<.05$; $OR=0.32$). As one Reading resident put it, "if you're over-prepared, it looks like you're [at high risk]. Even though the reality might be that actually you're prepared [and] therefore flooding wouldn't matter, it would still put [people] off [buying your property]".

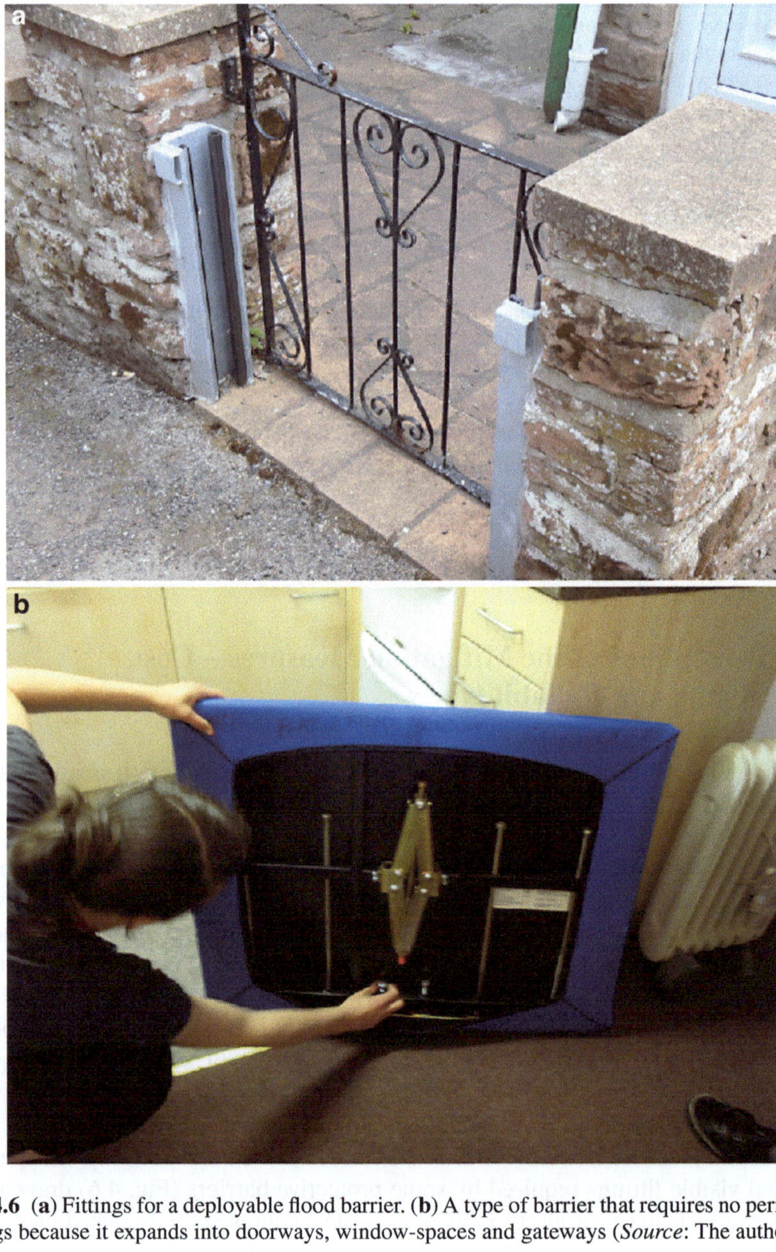

Fig. 4.6 (**a**) Fittings for a deployable flood barrier. (**b**) A type of barrier that requires no permanent fittings because it expands into doorways, window-spaces and gateways (*Source*: The author)

A third practical issue concerns the perceived reliability of mitigation measures. Unreliable measures increase anxiety and the emotional impact of an environmentally destructive event is exacerbated when human error (such as the choice of a wrong measure) is seen to have contributed to the destruction (Brown et al. 2005).

In Survey 1, 27% of respondents agreed with the statement, "I don't think I'd be able to choose the right way to protect my home"[2] and participants in the qualitative interviews expressed concern that they would feel or look foolish if it later became evident that they had made wrong choices or been 'ripped-off' by those selling flood protection products.

In the UK, the significance of these three types of barriers is gradually reducing. Some local authorities and charities offer support with the costs of mitigation measures; insurance companies sometimes agree to pay for flooded homes to be adapted after a flood; newer mitigation measures are more subtle in appearance, and a small number of independent experts are willing to give advice on the selection of measures. For the time being, however, there is little public awareness of these developments and their use remains patchy.

4.4 Anxiety, Fatalism and Vulnerability

Furthermore, qualitative analysis (Harries 2008b) suggests that these more overt justifications for not mitigating flood risk are not necessarily the most influential. The more important motivating factor, it suggests, might be the desire to protect those core elements of social representations that are important for what Giddens (1991) calls *ontological security* – the feeling of existential safety and meaningfulness that results from a belief in the continuity of one's identity and existence.

Social representations theory posits that people's representations of the world consist of networks of peripheral elements that cluster round a central core (Abric 1984; Bangerter 2000). Any threat to the core elements of this representational system causes profound anxiety (Wilkinson 2001) and threatens ontological security, so unfamiliar information, concepts and experiences are normally assimilated in such a way as to prevent any impact on the core and are only allowed to affect outer layers of the representational network.

The response to new information about flood risk typifies this phenomenon. To accept the existence of a flood risk would be to jeopardise core elements of three social representations that are fundamental to ontological security. It would be to accommodate within the representation of 'home', the concept of danger; to include in the representation of 'society', the idea that it is sometimes incapable of protecting its citizens, and to represent 'nature' as a source of threat as well as of pleasure and satisfaction.

One way of protecting core representational elements is to employ fatalism and blame as means of attributing flood risk to external factors rather than to intrinsic characteristics of nature, home and society. Fatalism implies that events are governed by chance rather than by any kind of pattern and allows people to

---

[2] Agreement with this statement was not statistically associated with reported use of mitigation or protection measures.

consider the likelihood of their experiencing a flood as identical to that of others. If their home is flooded, this is considered as 'bad luck', an 'act of God' or a 'freak event', but not as indicative of any pattern. Similarly, attributing blame to an external individual implies that the situation can be ameliorated once that person changes their behaviour and that the state of risk is not inherent to the core essence of 'home', 'nature' or 'society'. Blame and fatalism are not compatible with the discourse of mitigation and lead, therefore, to a reluctance to take practical risk reduction steps.

Qualitative analysis of interview data suggests that blame and fatalism discourses are associated with lack of access to the skills necessary for making practical adaptations to flood risk – i.e. the ability to understand and intervene in the relationship between human-made structures and water. Those with occupational backgrounds that furnish them with the relevant abilities and confidence – e.g. engineering or farming – are more able to make practical adaptations and therefore have less need of blame and fatalism; and the same is true for those whose social networks include such people. For example, in the Cumbrian town of Appleby, building and agriculture were such common professions that most residents either had the relevant skills themselves or could access them via social networks. Here, there was little evidence of fatalism. In London, by contrast, social networks were more widely distributed, the sharing of practical skills was less viable and most residents worked in professions not consonant with the skills relevant to practical flood risk management. Here, fatalism and blame were more evident.

4.4.1 The Representation of 'Home'

The first of the social representations that is protected in this manner is that of 'home'. In Western societies, 'home' generally elicits notions of continuity, safety, relaxation, privacy and familial affection (Cooper 1976; Mallett 2004; Saunders 1989; Smith 1994) – even where, as is often the case (e.g. Mallett 2004), this does not reflect lived experience. As people spend time in their homes, the routines they develop there, the aspects of their identities that they project onto the building fabric and the accretion of personal and inter-personal memories all imbue the place with their sense of who they are (Tuan 1974). Indeed, 'home' can be seen as a fixed and sacred spot from which people can create a version of the universe that fits in with their desires (see Cooper 1976). Any invasion of the home therefore undermines ontological security (Dupuis and Thorns 1998) and the temporary loss of home due to forced relocation after flooding is associated with deterioration in mental health (Ohl and Tapsell 2000; Tapsell et al. 1999).

The importance of this representation reveals itself in a reluctance to consider mitigation measures that would lessen the conformity of people's homes to the idealised norm. This is illustrated by an interview with a professional woman who, 2 years previously, had returned to find that a sudden flood had washed through the ground floor of her home.

Interviewer	If you were able to do things you could just leave in place and forget about… I don't know what that might be. It might be… raising your doorway for example your floor a little bit taking some measure permanently. Would that be better?
Martha	[…] I think we don't really want to (pause) change it – I like my house to look nice – I don't want to have a door that is like a bit daft because I raise the (laugh). And each time when we have friends or people coming through, you say well, you know, 'can you please step higher' (laughter).

A second excerpt is taken from a focus group of working class respondents:

Interviewer	Someone I spoke to […] he got this big whacking board that he can screw in across his front door […]
Marc	Yeah but then again, you'd feel like a prisoner (laughter)—a prisoner in your own home init (laughter)!
Pat	Yeah, prisoner in your own home!
Marc	Prisoner in your own home (laughter) […] you might get squatters moving in while you're out! (Laughter)
Freddy	The trouble is, you've got no flood coming in, but then a fire starts and you've had it! (Laughter)

Here, the respondents defend the concept of the home as a place of conformity to norms and represent it as a place of safety and comfort. The idea of barring the gateway between home and the rest of the world is interpreted as restricting freedom ("prisoner in your own home"), and inviting invasion ("you might get squatters") and danger ("then a fire starts and you've had it!"). The laughter and hyperbole in this excerpt seem to be an example of what Konrad Lorenz calls 'a controlled form of aggression' (cited in Morreall 1983, p. 6). Laughter appears to be employed as a means to ridicule invasive, alien representations that might threaten respondents' own representations of their homes as places of safety; and the use of hyperbole enhances the opportunities for ridicule by exaggerating the incongruity between the idealised concept of home and the alien concepts being mooted.

4.4.2 The Representation of 'Society'

The representation of society, too, plays a role in protecting people's ontological security against natural hazards. If they can represent society as essentially just and effective, as providing rescue and recompense or as preventing destructive events from ever occurring, then they can more easily represent their homes as safe and feel ontologically secure. In most of the Western world, society is represented as one of the main guarantors of the security of people's homes. Somewhat counter-intuitively, the frequent use of the discourse of blame amongst respondents suggests that they are attempting to protect that notion, for blaming a body implies that it retains the capacity to behave otherwise and even that (in the normal course of things) it *should* behave otherwise.

4.4.3 The Representation of 'Nature'

A final pillar of support for the phenomenological safety of home is the representation of nature as benign. Although increasingly challenged by the climate change discourse, this representation has been a dominant one in the West, where nature is often represented as a realm of positive moral influence (Macnaghten and Urry 1995; Soper 1995) and the destructive aspects of its character are not usually treated as part of the normal spectrum of human-environment relations (see Hewitt 1983). Hence, in spite of being aware of natural disasters such as the 2004 Boxing Day tsunami in Asia and the recent flash flood in Boscastle, most respondents still represented the role of nature in their own lives as essentially positive. As the following passage illustrates, such a representation makes flooding seem less threatening. Although these respondents' cottage is regularly threatened by flooding, the comparison with burglary indicates a representation of natural events as relatively benign ("It's a natural phenomenon, isn't it"):

George	I'd sooner have water [than burglary] I think.
Interviewer	How comes?
George	It's a natural phenomenon, isn't it.
Margaret	You can't help that.
George	Water, to me, it's natural—apart from all the buildings created it—you might say.
Interviewer	Yeah, yeah.
George	It's a normal…natural phenomenon, I think—flooding. It's from rain and flood, isn't it. Act of God, you could…Would that just about cover it? […]

George's suggestion that burglary is less acceptable than flooding because of the presence of intent and malice is echoed by other respondents. As one indicates, floods that are attributed to people are more detrimental to the feeling of security than floods that are attributed to nature (which, in the following passage, is equated with 'God'):

Florence	Then they started to think about regulating the flooding and opening and shutting the doors of the Thames and that; and I must say that since then, I personally have felt that it was no longer an act of God which was happening, but controlled by the powers that be. In other words, the last [time…] we felt that whoever it was had decided to flood us rather than flood the centre of Reading. So my perception is now—from fatalistic, before: 'floods will happen; the river is a risk; we're ready to take it'… […] I would say that over the last 10 years I've become a bit cynical, in the sense that I felt much more regulated by a central flooding control, which means that if they decide to flood us, they will. […] And we heard it said that they decided not to flood the centre of [name of their town] because there was electricity generators [there] and therefore we've … I felt a lot more insecure.

Florence's construction of the causality of floods, like George's construction of the nature of floods, limits damage to ontological security by blaming the more threatening floods on human intervention and thereby preserving the representation of nature as benign. Flood risk is represented as a threat to ontological security only as long as people continue their malign interference and if left to behave 'naturally' Florence's home would suffer no adverse effects. Seen from within this representational framework, it is human interventions that disturb the natural system, and Florence's home is not inherently at risk because the situation will change as soon as humankind begins to behave differently.

This tendency to attribute more 'risky' floods to people rather than to nature is evident throughout the sample and is independent of the type of respondent or the source of the floods: rivers, groundwater, sewers etc. Rather than resulting from a perception of what is the material cause of a flood, the attribution of blame is a convenient tool for the protection of a particular representation of nature and of the security that this representation allows.

Blame and fatalism are added to the representational network in order to protect the inner core of representations designed to avoid anxiety. However, although they are instrumentally useful for the protection of essential feelings of security, they also de-legitimise discourses that promote local adaptation. In other words, there is a measure of incompatibility between ontological security and the implementation of practical measures to ensure the practical protection of health and property.

4.5 Identity, Trust and Responsibility

Blame de-legitimises the adaptation discourse not only at the individual level but also at the level of society. This is because it is antithetical to two of the essential components of adaptive behaviour: trust and responsibility. To help explain this, a theory of social identity is now introduced to the reader.

Social identity theory (Abrams and Hogg 1990; Hogg and Abrams 1988; Tajfel 1982; Turner 1982, 1985) examines the role that groups play in determining the thoughts, feelings and behaviours of their individual members. SIT theorists argue that categorisation is essential for the creation of understanding and identity (Tajfel 1972) and that successful self-categorisation as a group member is a necessary part of functional success in the social world.

Self-categorisation, it is asserted, prompts social comparisons of the self with other members of the group, leading to pressure for conformity of thoughts, feelings and actions. People rely on those with similar categorisations to themselves for both information about social reality and for approval of their beliefs, feelings and behaviours. This produces a tendency to conform to what is known as the group-prototype – the characteristics and behaviours of the notional person who embodies the group's core ideals.

This is not a deterministic process. Individuals choose the category of people with whom to identify and how much they wish to conform to the norms of that group. According to social identity theory, the outcomes of this first choice are context-dependent – people's identities are multi-faceted and the facet they choose to emphasise at any time depends on the situation. On each occasion, it is argued, they choose the dimensions of identity that are most salient to the current circumstances and then select the social category – and social identity – that provides the closest fit along those dimensions (Hogg and Abrams 1988). When the issue in question is flood risk, the most salient social dimension is exposure to the flood risk or experience of flooding. As a result, even where no pre-existing identity groups match the catchment of a flood risk, such a group will normally be formed during and immediately after a flood event. This is illustrated in the following excerpt from an interview with a resident of Reading who had experienced relatively minor flooding:

> Everybody was there; we were all involved in the same process of deciding whether the middle pathway was going to be closed – whether the postman would deliver or if we had to go and pick up our post. As if we were becoming a kind of a little community which was surviving an act ... you know, an act of God

This quote illustrates the nascent sense of belonging that can be generated by the experience of flooding. A social identity focused on residents' experiences is suggested by the stress on collective action ("we were all involved in the same process"), the affectionate use of the diminutive noun phrase "little community" and the association of 'community' with 'survival' ("we were becoming a kind of little community which was surviving").

After the excitement of the flood itself, however, the focus of the social identity group often turns from survival to questions of blame; and it is to members of *out-groups* that blame is usually attributed. Social identity theory maintains that in order to protect group identity and the benefits that it brings, groups maximise the positive difference between their representations of themselves (the *in-group*) and their representations of other groups (*out-groups*). Hence, they accentuate positive characteristics of the in-group and attenuate their negative characteristics; and they do the reverse for out-groups. As a result, when there has been a flood, victims tend to idealise the qualities of the in-group that comprises themselves and fellow victims and to attribute negative qualities, including blame, to members of out-groups such as government agencies and local authorities.

This has numerous consequences for the behavioural response to the risk of future flooding. Attempts to emphasise the demarcation between *in-groups* and professional flood risk managers inevitably reduce the possibility of successful information sharing and mutual assistance. Environment Agency staff often experience people from flooded areas as hostile and aggressive. When they respond by distancing themselves, either emotionally or by avoiding contact, this confirms the prejudices of the flood victims, who now see themselves as faced with a group of professionals who cannot relate to them and seem insensitive to their suffering or anxieties. Although the intensity of these responses reduces as memories of the flood event become more distant, the nature of the relationship between residents and professionals will often have been set.

This, in turn, has implications for views on who is responsible for managing the flood risk. As long as blame is attributed to outsiders, so too is responsibility for remedying the situation; for if flood victims were to accept responsibility themselves, this would raise the question of why they did not do so before the flood and would, therefore, weaken the positive representation of in-group qualities. This assertion of in-group identity can also be seen as a reaction against the rising trend towards individualisation (Bickerstaff and Walker 2002). Under advanced liberalism, governments increasingly seek to influence as well as to command, and use the discourse of individual responsibility as an alternative means to control populations (Raco and Imrie 2000). What is sometimes interpreted as a strategy to avoid changing habitual behaviour and save decision costs (Lindbladh and Lyttkens 2002) can therefore also be seen as a form of resistance against attempted control and, consequently, as a defence of social identity. Blaming public bodies is represented as a means of resisting the state, maintaining the boundaries between the in-group and the out-group and preserving identity. There is evidence of this amongst tenants of social housing in flood risk areas, where the suggestion of property-level flood resilience measures was in one instance represented as a contravention of human rights.

Geoff	That laminated flooring in my hall last year – I didn't think, 'oh I better not put this down because it might flood', I don't think that. […]
Rob	What about sofas, are we allowed them?
Interviewer	Yeah sofas, nothing [inaudible]…
Stuart	That's breaking human rights, init?
Jackie	That's comfort. They've got their comforts! Would they walk about on concrete floors? You just don't think that way.
Stuart	You don't see Tony Blair living on a concrete floor, do you!

When the maintenance of social identity precludes the acceptance of responsibility for flood risk, the implementation of mitigation measures by individuals becomes stigmatised and those that take such steps risk exclusion from the in-group. In one interview, the one resident on a street to have purchased a flood-board is represented as undermining local solidarity; and to preserve the positive identity of the in-group, he is spoken of with condescension and his actions disparaged. In the same interview, the prototypical in-group member who has made no preparations for flooding continues to be described as "very down-to-earth", "educated" and "practical".

An excerpt from a focus group also shows the assertion of in-group values in action:

Rob	You don't think, "I'll get this because this might happen". People just don't think that. You think, "I'll deal with it when it happens" […] If I met [a neighbour] today, putting sand in sandbags, and he said to me: "Just in case it floods", I'd be thinking, "he's nuts!" [*General laughter*]
Interviewer	How about if your neighbour was bolting [new floodgates] to his door?
Stuart	I'd ask him who's paying for it! […]
Jackie	"More money than sense", I would say!

In this discussion, participants use two rhetorical means to encourage compliance with group norms. They assert that it is not normal to plan for such eventualities ("people just don't think that way") and they ridicule such behaviour ("I'd be thinking, he's nuts!"; "More money than sense, I would say!").

Such social barriers to adaptive action disappear only when the relevant behaviours themselves are seen as typical characteristics of the in-group – i.e. either when they are adopted by in-group members who are seen to embody the characteristics of the proto-typical member (see Rogers' (1987) *diffusion of innovations theory*) or when the boundaries of the group are perceived to soften.

Evidence of both these phenomena was found in the case of the town of Appleby (see Harries 2010b). Here, staff from the Environment Agency and the district council approached the issue of property-level mitigation strategically by only approaching the question of property-level mitigation once historical differences between the authorities and the townspeople had been resolved and this out-group/in-group distinction much reduced. Rather than immediately trying to suggest a solution to the flood risk problem and confronting the residents with their failure to find a solution themselves, they offered to help townspeople implement their own solution (the more efficient distribution of sandbags), even though this was not perceived as very practically effective. This valorisation of local social identity helped undermine the negative representation of the flood risk professionals and weakened the demarcation between the townspeople and the professionals from outside the town. Consequently, when the more effective idea of property-level mitigation was mooted, it was readily accepted by residents of the town.

A key issue in this story is trust. The residents of Appleby had seen no evidence of the performance of either the mitigation measures being suggested nor of the flood risk management professionals – who heretofore had been seen to fail to adequately deal with the issue of flood risk in the area. As a result, there were few grounds for *calculative trust* – confidence based on evidence of past behaviour (Earle 2010; Rousseau et al. 1998).

In situations of acute flood risk, calculative trust will often be lacking. The desire to tackle a flood risk problem usually implies recent experience of flooding and this, in turn, implies that the problem has not yet been successfully tackled. Actors in the Appleby story overcame the absence of calculative trust by cultivating a substitute: *relational trust*. Relational trust stems not from experience of other people's actions but from an estimation of their values and intentions (Earle 2010) and it causes people to turn to friends and family for advice on risk issues before they turn to professionals (Rogers 1987). In Appleby, the professional flood risk managers developed relational trust by nurturing friendships with local people and convincing them that they shared a similar value-base and intentions. As a result, the operational distinction between out-group and in-group became blurred and the views and suggestions of the professionals became regarded with greater respect.

The existence of relational trust between two different social identity groups creates opportunities for new norms of behaviour. Where, previously, residents of Appleby might have seen the use of door-barriers as a vindication of the arguments of an antithetical group of outsiders, they now viewed it as a positive sign of how the town was working in partnership with a valued group of professionals who happened to come from outside the town. Furthermore, the presence of relational trust seems to have generated a desire to supplement the bond between town and the professional flood risk managers with evidence that the confidence of the former in the latter had been well-founded ("let's show them what we can do!"). Instead of denying responsibility in order to shift blame onto an out-group, the town took responsibility for managing the risk and, indeed, made this an additional source of pride for the town.

This suggests an interdependence between relational and calculative trust (see Earle 2010). In the long term, relational trust needs to be accompanied by evidence to justify calculative trust. In both Appleby and Leeds, although a sense of shared values and intentions facilitated a good relationship between residents and flood risk management professionals, it did not generate confidence in the effectiveness of the professionals' actions. An improved relationship had allowed the introduction of new behavioural norms, but these norms had yet to be fully tested. Residents remained sceptical of the effectiveness of the new norms and reserved judgement with regard to their new partners in flood risk management. In Leeds, residents had not experienced another flood since the installation of the mitigation measures and reported that they were only moderately less anxious than they had been before their installation. In Appleby, the good performance of the measures in a recent flood had reinforced the town's new relationship with the authorities, and residents placed greater confidence in their professional advice.

Identity, trust and responsibility are inextricably linked. The absence of trust creates identity demarcations that, in turn, incentivise the renunciation of responsibility. The examples of Appleby and Leeds suggest that the generation of relational trust between risk management professionals and at-risk communities can overcome some of the barriers to local adoption of responsibility and normalise the notion of mitigation in the at-risk community.

4.6 Event Experience and Frequency

A further critical ingredient in the recipe for successful urban adaptation is experience of the type of event in question. Numerous quantitative studies have shown experience to be a significant predictor of protective behaviour against natural hazards such as flooding (e.g. Grothmann and Reusswig 2006; Laska 1990; Siegrist and Gutscher 2008). This, it is argued here, is because experience of flooding transforms core social representations and the social identities that they support (Fig. 4.7).

In Survey 1 and Survey 3 there are significant relations between experience of flooding and the adoption of adaptive measures. Survey 1 reveals a strong relation between the use of flood barriers and the frequency of experience (χ^2 [3, $N=511$] $=57.92$, $p<.005$). This is independent of housing tenure, housing type or whether households had contents insurance (Table 4.2) and increases in strength with the number of floods experienced. People who had experienced one or two

Fig. 4.7 Flooded street in Appleby, 2009 (Karen Morley-Chesworth. Permission granted)

Table 4.2 Results of logistic regression onto the variable 'protective action taken' (Survey 1, $N=427$)

								95.0% C.I. for Exp(B)	
	N	B	S.E.	Wald	df	Sig.	Exp(B)	Lower	Upper
Occupants own/have a mortgage on the property	375	.274	.516	.282	1	.596	1.315	.478	3.618
Number of experiences of flooding in the home[a]		.770	.155	24.626	1	.000	2.160	1.594	2.928
Occupants have contents insurance	391	.386	.647	.355	1	.551	1.471	.414	5.228
Home has more than one floor (i.e. it is a house rather than a bungalow or flat)	336	.267	.344	.600	1	.439	1.305	.665	2.563
Constant	–	−2.760	.695	15.758	1	.000	.063	–	–

[a]Continuous variable

floods ($N=127$) were 3.92 times as likely to have taken such measures compared with those who had never been flooded ($N=368$), but those who had been flooded more than twice were 6.33 times as likely to have done so ($N=16$) (Table 4.3).

This finding is broadly confirmed by analysis of Survey 3, in which respondents were asked whether they had (1) obtained sandbags and sand, (2) installed pumps, flood-barriers or airbrick covers or (3) built new drains or protective walls. Once again there was a significant relationship between number of floods experienced and the use of barriers (χ^2 [3, $N=276$] = 10.82, $p<.05$) and this was independent of other available and salient variables: social class, housing tenure, age of respondent and

Table 4.3 Cross-tabulation of 'number of times flooded in the home' against 'protective action taken' (Survey 1)

		Protective measure taken by household?		
		No	Yes	Total
Number of times flooded in the home[a]	None	339	29	368
	One	57	29	86
	Two	30	11	41
	More than two	8	8	16
	Total	434	77	511

[a]The continuous variable in the original dataset has here been converted into a categorical variable

Table 4.4 Results of logistic regression onto the variable 'protective action taken' (Survey 3, $N=266$)

	N	B	S.E.	Wald	df	Sig.	Exp(B)	95.0% C.I. for Exp(B) Lower	Upper
Occupants own/have a mortgage on the property	243	.146	.490	.089	1	.766	1.16	.443	3.023
Number of experiences of flooding[a]	–	.379	.122	9.61	1	.002	1.46	1.15	1.855
Social grades C2, D or E	104	−.323	.275	1.38	1	.239	.724	.423	1.240
Costs incurred by most recent flood were not covered by insurance	156	−.248	.262	.891	1	.345	.781	.467	1.306
Age – 18–34	19	–	–	1.89	2	.389	–	–	–
Age – 35–54	94	.568	.565	1.01	1	.315	1.77	.583	5.346
Age – 55 and over	153	.727	.546	1.77	1	.183	2.07	.709	6.034
Constant	–	−1.178	.678	3.02	1	.082	.308	–	–

[a]Continuous variable

whether the most recent flood had incurred any net cost (Table 4.4). However, in this dataset experience of flooding only became significant when people had lived through at least three flood events. Respondents with experience of three or more floods were 1.67 times as likely to have taken protective measures as people who had never been flooded ($N=18$) and 1.59 times as likely as those who had only been flooded once ($N=171$) (Table 4.5).

One possible explanation for this phenomenon is that experience of multiple floods makes it more difficult for individuals to hold on to the social representations of 'nature', 'home' and 'society' that allow them to feel secure. It was argued above that people filter out evidence that might contradict these representations by depicting floods as 'freak' events or blaming others for their occurrence. The more, and the more vivid, the experiences of flooding, the harder it is to deny the evidence these experiences provide and the harder it is to protect core representations that depict home life as secure. The qualitative evidence suggests that when the pressure

Table 4.5 Cross-tabulation of 'number of times flooded' against 'protective action taken' (Survey 3)

		Protective measure taken by household?		
		No	Yes	Total
Number of times flooded in the home[a]	None	11	7	18
	One	101	70	171
	Two	14	16	30
	More than two	20	37	57
	Total	146	130	276

[a]The continuous variable in the original dataset has here been converted into a categorical variable

to revise these representations becomes too great, people either fall into a state of anxiety or rely on a self-representation that depicts them as capable of dealing with the wconsequences of living in a less safe world.

4.6.1 When Existing Representations Have to Be Abandoned – The Stoical Response

Those who respond to multiple flood experiences in the latter manner normalise flooding and integrate the ongoing risk into their representations of everyday life. Rather than being perceived as an existential threat, flooding is defined as a threat to material security only. As a result, these householders no longer need to deny the risk and have need neither of the social representations described above, nor of the discourses of blame and fatalism that are often used to protect them.

This conclusion is supported by evidence from Survey 3 and the qualitative research. In Survey 3, respondents were asked whether they agreed with the statement "I prefer not to think about scary things like floods". Of those that had been flooded just once, 70% agreed with this statement ($N=169$). However, the figure dropped to 22% for those with two or more experiences of flooding ($N=87$). In the qualitative analysis, the more 'stoical' participants do not try to represent the threat of flooding as controllable or to represent themselves as able to neutralise the destructive effects of floods; nor do they represent life as innately safe, blame flooding on others or attribute it to 'bad luck'. Rather, they assert that they "accept…quite well" that flooding and flood risk are a "part of life"; that they are "philosophical" about flooding and do not get upset about it; that they are not "scared" of flooding like other people are, and that they are "tenacious" in the face of the risk. In other words, the risk of flooding is integrated into a representation that depicts it as normal for life to include losses as well as gains.

'Stoics' are able to describe the causes of flooding using a discourse that is predominantly technical and with little evident emotionality. The risk is not denied, but neither does it provoke anxiety. This, as the following excerpt from the focus group with Reading professionals illustrates, allows everyday life to go on.

Craig	[…] I'm old enough and long enough in the tooth to realise that a bit of wet carpet and a little bit of re-decorating, actually in the overall scheme of life, isn't that important. Um, and there are other things that are much more important. And therefore if it costs a couple of grand – even out of my own pocket – to replace a fridge, a freezer, some carpet and a bit of kitchen, which I might want to change anyway
Christopher	You're hoping it will flood, really, aren't you! [*Laugh*]
Craig	You know, how incredibly important is that? You do weigh that up against the hassle of moving, the cost of moving, the fact that you like where you live and so on.
Interviewer	But yet there *is* worry. You *are* worried about water coming in. Even though, yes, on the one hand you're saying it's only possessions and it would only be, like, a bit of re-decorating; but on the other hand it is a cause of concern, isn't it?
Craig	Yes, but it doesn't fill my every waking moment.
Joan	No [*Laugh*].
Craig	And that's it; at the end of the day, it is a *concern*.

Craig seems to want to present himself as not worried about the flood risk. Although he admitted earlier in this interview that he experienced some fear when floodwaters were about to enter his home, it is clear from the final sentence of this excerpt that he prefers to be seen as concerned rather than fearful ("at the end of the day, it is a *concern*"). Furthermore there is no sense in Craig's talk that flooding threatens anything other than his material possessions, for he stresses that his home has "no particular sentimental value". His language, too, implies a rational appraisal of the risk and not an emotional one (i.e. "weigh[ing] up"; "costs"; "hassles" and "facts"). In keeping with other stoics in the sample, flood risk does not seem to undermine Craig's ontological security. His fear that his home seemed about to be flooded has not overflowed into his everyday life, so there is no long-term anxiety about possible future flooding.

4.6.2 The Response When Existing Representations Have to Be Abandoned – Anxiety

Experiences of flooding only seem to reduce anxiety, however, if they are not too traumatic. Rather than being stoical, flood victims that are traumatised by their experiences of flooding enter an emotional crisis. No longer being able to sustain the representations of 'nature', 'society', 'home' and 'self' that protected their ontological security, they fall into a state of anxious insecurity.

In Survey 2, a series of questions designed to measure Post Traumatic Stress Disorder (PTSD) was used as a proxy for the loss of ontological security. PTSD is indicated by the re-living of the traumatic event, a numbing of general responsiveness and persistent symptoms of general arousal (e.g. irritability, difficulty sleeping and lack of concentration) (American Psychiatric Association, 1994, as cited in Joseph et al. 1997). It was measured in the survey using a version of the Post Traumatic Stress Scale that had been adapted to relate specifically to experience of household flooding (Dua and Scott 2001). In the dataset, 2% of flooded individuals had 'high' or 'extreme' PTSD scores that suggested the loss of ontological security.

There were two apparent examples of anxious insecurity in the qualitative sample. One respondent, a labourer, says that his flat is the "first stable home" he has had and that he would "go crazy" if it were flooded again. This suggests heavy dependency on 'home' as a place of safety and stability and also that this representation is vulnerable to challenge. Furthermore, the respondent shows faith neither in his own ability to mitigate the risk of another flood nor in that of the local authority ("the council wouldn't even put down a bit of grit on the road if it snowed, never mind spend money on [flood risk alleviation]").

A second example of anxious insecurity is a professional who had twice been evacuated as a result of a flood and whose home had on both occasions been badly damaged. She represents 'nature' as callous and destructive and, after the failure of collective local action against the local authority, despairs of receiving any help from the state. Having seen her home stripped of all its homeliness ("what was a home [...] suddenly just becomes bricks"), she seems to have abandoned the representation of her flat as a safe centre for her life and identity and to have no faith in household-level mitigation measures or her own ability to protect her home ("there's nothing you can do"; "all the nuts and bolts and sandbags are not really...are not going to solve this"). As a result she too reports a lack of emotional security:

Interviewer How does it feel during the summer when you are aware that there could be flooding around? Some people have told me that there's a kind of anxiety involved...

Vikki I get hysterical, absolutely hysterical. This last time, when we didn't get flooded, I found I was getting really in a state about things. I was getting panicky and I was on the internet every single night looking at the weather forecast and going into all the details. And then this flood warning thing you can look up as well; and it was ridiculous, I found myself doing it every single day and I was a nervous wreck [...] I get hysterical when it happens. I start shaking and I can't speak, it's almost like I'm in shock.

Patterns in the qualitative data suggest that the severity of the flood experience might be one important predictor of this loss of ontological security – particularly the speed at which floodwaters rise, perceived pollutant levels and the extent of the actual or potential damage and disruption. This finds some support in previous analysis of the Survey 2 data by Tunstall et al. (2006), who found that high PTSD scores amongst flooded householders correlated with, *inter alia*, the experience of evacuation, the depth of the flooding in the main rooms of the home, the time it took to return to normality and the perceived contamination of the floodwaters.

4.7 Conclusion

For many householders in the UK, the perceived costs of taking adaptive measures against flood risk are greater than the anticipated gains. UK citizens have become habituated to the idea that the state will protect them from natural hazards such as flooding. As a result, the representation of 'home' as a place of innate safety continues to form a key pillar of ontological security for many people and the notion of property-level mitigation has yet to be assimilated into the corpus of 'normal' behaviours. At the same time, the reliability of such measures and the trustworthiness of their proponents are often doubted by members of the public. As a result, the risk involved in implementing mitigation measures is sometimes seen as greater than the reduction in risk that they will bring. Although it would be reductionist to assume that people use any such rational form of cost-benefit analysis to reach their decisions, this argument makes it clear that the decision not to adopt practical adaptations can be seen as just as instrumental as the decision to adopt them.

Relatively high levels of take-up in the areas benefitting from Defra's grant scheme suggest that some of these barriers can be overcome. Not only did the scheme pay for the purchase and implementation of measures; it also provided expert advice on what were appropriate measures. Doubts over which measure to take exacerbate anxiety about regret and increase the likelihood of inaction (see Zeelenberg et al. 2002), especially where, as is the case with flood risk, inaction is the norm (see Tykocinski and Pittman 1998). The UK police and fire services already offer expert, individually tailored guidance on the prevention of burglaries and fires, and there is a need for similar independent advice to be made available for situations of flood risk.

A further advantage of the Defra scheme was its engagement of householders at the collective level. In those areas that participated in the scheme, no single householder had to take the exposing step of being the first to install a door-board, airbrick cover etc. Communities were approached collectively, so individuals were able to reduce their risk of blame or stigma by sheltering behind the decisions of the group. Furthermore, research in the pilot scheme areas in Appleby and Leeds suggests that where local authorities and the Environment Agency were able to foster a relationship of partnership with at-risk groups, the boundary between in-group (the flood victims) and out-group (the authorities) was weakened (Harries 2009, 2010b). Where this occurs, residents are less able to use the strategy of blame and fatalism and will be more amenable to the notion of property-level mitigation.

However, experience of flooding is also critical to the take-up of mitigation measures. The government grant scheme only applied to areas where the probability of flooding was particularly high and there had been repeated recent flooding. Although it is not possible to gauge what the scheme's impact would have been on less flood-prone areas, the evidence presented in this chapter indicates that it would have met with far less success. Experience, it has been suggested here, wears down the defences that people use to protect the representational structures that allow them to feel secure. If the representations that replace the old, discredited ones include the idea that natural hazards can be survived, then denial is no longer an instrumentally functional strategy and physical adaptations become more emotionally viable.

This chapter has made an argument for greater consideration of the role of *anticipated emotions* – feelings that people believe they will feel in hypothetical future scenarios (Bagozzi et al. 2000), particularly regret arising from counterfactual comparisons (Loomes and Sugden 1982). Although researchers have become increasingly aware of the significance of emotions in determining risk response (Harries 2012; Slovic 2000; Slovic et al. 2004), practitioners too often assume that perceptions are the result of the intellect alone. To avoid this error, it is necessary to look beyond people's *post hoc* justifications for taking or not taking adaptive steps and to try to understand the latent drivers of behaviour. This chapter has attempted to do this. It suggests that before they will act to protect themselves, their families and their property, people not only need information and the necessary financial resources, but also reassurance about the implications of protective action for their emotional security.

References

Abrams D, Hogg M (1990) An introduction to the social identity approach. Harvester Wheatsheaf, Hertfordshire

Abric JC (1984) A theoretical and experimental approach to the study of social representations in a situation of interaction. In: Farr R, Moscovici S (eds) Social representations. Cambridge University Press, Cambridge, pp 169–183

Austin J (1962) How to do things with words. Oxford University Press, Oxford

Bagozzi RP, Baumgartner H, Pieters R, Zeelenberg M (2000) The role of emotions in goal-directed behavior. In: Ratneshwar S, Mick DG, Huffman C (eds) The why of consumption: contemporary perspectives on consumer motives, goals, and desires. Routledge, New York, pp 36–58

Bangerter A (2000) Transformation between scientific and social representations of conception: the method of serial reproduction. Br J Soc Psychol 39(4):521–535

Bickerstaff K, Walker G (2002) Risk, responsibility and blame: an analysis of vocabularies of motive in air-pollution(ing) discourses. Environ Plan A 34(12):2175–2192

Bohrnstedt GW, Knoke D (1984) Statistics for social data analysis, 3rd edn. F E Peacock Publishers, Itasca

Brown TC, Petersen GL, Brodersen MR, Ford V, Bell PA (2005) The judged seriousness of an environmental loss is a matter of what caused it. J Environ Psychol 25(1):13–21

Cooper C (1976) The house as symbol of the self. In: Proshansky HH, Ittelson WH, Rivlin LG (eds) Environmental psychology: people and their settings, 2nd edn. Hold, Rinehart and Winston, New York, pp 435–448

Department for Environment, Food and Rural Affairs (2005) Making space for water: taking forward a new government strategy for flood and coastal erosion risk management in England. Defra Publications, London

Department for Environment, Food and Rural Affairs (2008a) Consultation for policy options for promoting property-level flood protection and resilience. Defra, London

Department for Environment, Food and Rural Affairs (2008b) Resilience grants pilot projects: executive summary. Defra, London

Dua J, Scott W (2001) Manual for the post-traumatic stress scale. Psych Press, Melbourne

Dupuis A, Thorns DC (1998) Home, home ownership and the search for ontological security. Sociol Rev 46(1):24–47

Earle TC (2010) Trust in risk management: a model-based review of empirical research. Risk Anal 30(4):541–574

Evans E, Ashley R, Hall J, Penning-Rowsell E, Saul A, Sayers P, Thorne C, Watkinson A (2004) Foresight, future flooding. scientific summary, vol I, Future risks and their drivers. Office of Science and Technology, London

Fairclough N (2003) Analysing discourse: textual analysis for social research. Routledge, London

Giddens A (1991) Modernity and self-identity. Polity Press, Cambridge

Greater London Authority (2010) Proposal for an investigation into rain and river flood risk management in London. Executive Director of Secretariat, Greater London Authority. http://legacy.london.gov.uk/assembly/envmtgs/2010/jul15/item10.pdf. Accessed 20 Sept 2010

Grothmann T, Reusswig F (2006) People at risk of flooding: why some residents take precautionary action while others do not. Nat Hazard 38(1–2):101–120

Guardian (1998) The flood of the century. The Guardian, 13 Apr 1998, p. 8

Halliday M (1994) An introduction to functional grammar. Edward Arnold, London

Harries T (2008a) Householder responses to flood risk: the consequences of the search for ontological security. Doctoral dissertation. Flood Hazard Research Centre, Middlesex University, Enfield

Harries T (2008b) Feeling secure or being secure? Why it can seem sensible not to protect yourself against a natural hazard. Health Risk Soc 10(5):479–490

Harries T (2009) Understanding barriers and incentives for Retrofitted household flood protection and resilience (Joint Defra/EA Flood and Coastal Erosion Risk Management R&D Programme). Defra, London

Harries T (2010a) Summary of findings: national flood forum 2010 survey of flooded individuals. National Flood Forum. http://floodforum.org.uk/files/Summary%20of%20findings.pdf. Accessed 20 Sept 2010

Harries T (2010b) Review of the pilot flood protection grant scheme in a recently flooded area (Joint Defra/EA Flood and Coastal Erosion Risk Management R&D Programme). Defra, London

Harries T (2012) The anticipated emotional consequences of adaptive behaviour: impacts on the take-up of household flood-protection protective measures. Environ Plan A 44(3):649–668

Hewitt K (1983) The idea of calamity in a technocratic age. In: Hewitt K (ed) Interpretations of calamity: from the viewpoint of human ecology. Allen & Unwin, London, pp 4–32

Hogg M, Abrams D (1988) Social identification: a social psychology of intergroup relations and group processes. Routledge, London

Johnson C, Tunstall S, Penning-Rowsell E (2005) Floods as catalysts for policy change: historical lessons from England and Wales. Int J Water Res Dev 21(4):561–575

Johnson C, Tunstall S, Penning-Rowsell E (2008) Crises as catalysts for adaptation: human response to major floods report for the ESRC environment and human behaviour new opportunities programme. Flood Hazard Research Centre, London

Joseph S, Williams R, Yule W (1997) Understanding post-traumatic stress: a psychosocial perspective on PTSD and treatment. Wiley, Chichester

Laska S (1990) Homeowner adaptation to flooding: an application of the general hazards coping theory. Environ Behav 22(3):320–357

Lindbladh E, Lyttkens CH (2002) Habit versus choice: the process of decision-making in health-related behaviour. Soc Sci Med 55(3):451–465

London Assembly (2002) Flooding in London: a London assembly scrutiny report. http://legacy.london.gov.uk/assembly/reports/environment/flooding.pdf. Accessed 20 Sept 2010

Loomes G, Sugden R (1982) Regret theory: an alternative theory of rational choice under uncertainty. Econ J 92(368):805–824

Macnaghten P, Urry J (1995) Towards a sociology of nature. Sociology 29(2):203–220

Mallett S (2004) Understanding home: a critical review of the literature. Sociol Rev 52(1):62–89

Market Research Society (2002) Occupation groupings: a job dictionary, 5th edn. The Market Research Society, London

McCarthy JJ, Canziani OF, Leary NA, Dokken DJ, White KS (2001) Climate change 2001: impacts. Cambridge University Press, Cambridge

Morreall J (1983) Taking laughter seriously. SUNY, Albany
Observer (2007) After the deluge, Britain will be swamped by a £6bn tidal wave of costs. http://observer.guardian.co.uk/business/story/0,,2136756,00.html. Accessed 30 July 2007
Ohl CA, Tapsell SM (2000) Flooding and human health: the dangers posed are not always obvious. Br Med J 321(11):1167–1168
Pitt M (2008) Learning lessons from the 2007 floods: an independent review by Sir Michael Pitt. Cabinet Office, London
Potter J, Wetherell M (1987) Discourse and social psychology: beyond attitudes and behaviour. Sage, London
Raco M, Imrie R (2000) Governmentality and rights and responsibilities in urban policy. Environ Plan A 32(12):2187–2204
Rogers EM (1987) The diffusions of innovations perspective. In: Weinstein ND (ed) Understanding and encouraging self-protective behaviour. Cambridge University Press, Cambridge, pp 79–94
Rousseau DM, Sitkin SB, Burt RS, Camerer C (1998) Not so different after all: a cross-discipline view of trust. Acad Manage Rev 23(3):344–354
RPA, EFTEC, CASPAR and Flood Hazard Research Centre (2004) The appraisal of human-related intangible impacts of flooding. Defra, London
Saunders P (1989) The meaning of "home" in contemporary English culture. Hous Stud 4(3):177–192
Siegrist M, Gutscher H (2008) Natural hazards and motivation for mitigation behavior: people cannot predict the affect evoked by a severe flood. Risk Anal 28(3):771–778
Slovic P (2000) Introduction and overview. In: Slovic P (ed) The perception of risk. Earthscan, London, pp 21–37
Slovic P, Finucane M, Peters M, Macgregor DG (2004) Risk as analysis and risk as feelings: some thoughts about affect, reason, risk, and rationality. Risk Anal 24(2):311–322
Smith SG (1994) The essential qualities of the home. J Environ Psychol 14(1):31–46
Smith K, Ward R (1998) Floods: physical processes and human impacts. Wiley, Chichester
Soper K (1995) What is nature? Culture, politics and the non-human. Blackwell, Oxford
Tajfel H (1972) La categorisation sociale. In: Moscovici S (ed) Introduction á la Psychologie Sociale, vol 1. Larousse, Paris
Tajfel H (1982) Social identity and intergroup relations. Cambridge University Press, Cambridge
Tapsell S, Tunstall SM, Penning-Rowsell EC, Handmer JW (1999) The health effects of the 1998 easter flooding in Banbury and Kidlington. Flood Hazard Research Centre, Enfield
Thurston N, Finlinson B, Breakspear R, Williams N, Shaw J, Chatterton J (2008) Developing the evidence base for flood resilience. R&D Technical Report FD2607/TR. Defra, London
Tuan YF (1974) Space and place: humanistic perspective. Prog Geogr 66(2):233–246
Tunstall S, Tapsell S, Green C, Floyd P, George C (2006) The health effects of flooding: social research results from England and Wales. J Water Health 4(3):365–380
Turner JC (1982) Towards a cognitive redefinition of the social group. In: Tajfel H (ed) Social identity and intergroup relations. Cambridge University Press, Cambridge
Turner JC (1985) Social categorization and the self-concept: a social cognitive theory of group behaviour. In: Lawler EJ (ed) Advances in group processes: theory and research, vol 2. JAI Press, Greenwich
Tykocinski O, Pittman T (1998) The consequences of doing nothing: inaction inertia as avoidance of anticipated counterfactual regret. J Pers Soc Psychol 75(3):607–616
Wilkinson I (2001) Anxiety in a risk society. Routledge, London
Wittgenstein L (1958) Philosophical investigations. Basil Blackwell, Oxford
World Health Organisation (2002) Floods: climate change and adaptation strategies for human health, meeting report. WHO Regional Office for Europe, Copenhagen
Zeelenberg M, van den Bos K, van Dijk E, Pieters R (2002) The inaction effect in the psychology of regret. J Pers Soc Psychol 82(3):314–327

Chapter 5
A Historical Overview of Social Representation of Earthquake Risk in Japan: Fatalism, Social Reform, Scientific Control and Collaborative Risk Management

Katsuya Yamori

Abstract The historical development of social representations of earthquake risk in Japan can be summarised in terms of four stages: *fatalism* (before the eighteenth century), *social reform* (from the mid-nineteenth century), *scientific control* (after World War II) and *collaborative risk management* (after the Great Hanshin-Awaji Earthquake in 1995). This evolution is not a linear, chronological movement through these modes of representation. Rather, these four basic styles of risk representation create a multilayered structure, allowing two or more types of risk representation to coexist. In Japanese society since the 2011 Great East Japan Earthquake, in particular, these four types of social representation of earthquake risk have developed into a complicated mixture. Seismologists have publicly referred to the 'failure of earthquake sciences', and criticism from society is increasing in regard to the ineffectiveness of safety measures regarding nuclear power plants. Thus, confidence in scientific control, which previously had been trusted and relied upon, is greatly shaken. Concurrent with this declining trust in science, vernacular science knowledge regarding disaster and disaster prevention is being disseminated at much higher rates to society at large, and the line separating such vernacular knowledge, older traditional views and 'pure' scientific knowledge is being blurred. Hence, a growing trend towards cognitive polyphasia (Moscovici 1976) is apparent. Furthermore, because widespread damage occurred in spite of the mutual cooperation of residents of local communities at the time of the emergency tsunami evacuation, the limits of collaborative risk management, which came into prominence after the Great Hanshin-Awaji Earthquake, have been quickly indicated. Meanwhile, through statements such as that by a prominent politician claiming that "this disaster is divine punishment for our selfishness", and arguments that the Great East Japan Earthquake of 2011 should represent a great

K. Yamori (✉)
Disaster Prevention Research Institute, Kyoto University,
Gokasho, Uji, Kyoto 611-011, Japan
e-mail: yamori@drs.dpri.kyoto-u.ac.jp

H. Joffe et al. (eds.), *Cities at Risk: Living with Perils in the 21st Century*,
Advances in Natural and Technological Hazards Research 33,
DOI 10.1007/978-94-007-6184-1_5,

turning point for an economically and politically stagnant Japanese society, it has become evident that old risk representation styles such as fatalism and social reform also remain firmly entrenched.

5.1 Cities in Japan at Risk from Earthquakes

5.1.1 The 1995 Great Hanshin-Awaji Earthquake

The 1995 Great Hanshin-Awaji (or Kobe) Earthquake killed over 6,400 people. This was an unnerving incident for the Japanese people, who came to realise that many cities in Japan are at high risk of earthquake damage (Fig. 5.1). In reality, Kobe (population 1.2 million) and other western Japanese cities were less prepared for earthquakes than were cities such as Tokyo, Nagoya, Shizuoka and Sendai in central or eastern Japan. This chapter suggests some reasons why Kobe was less prepared.

Firstly, Kobe had experienced few earthquakes in the twentieth century, whereas Tokyo and other cities in eastern Japan frequently experienced small- and medium-sized earthquakes. Additionally, whilst earthquake experts had warned that a large earthquake might hit Tokyo, Sendai, or another eastern city, few had considered Kobe as likely to experience a large event. Finally, prior to the 1995 earthquake, the damage and casualties resulting from earthquakes in Japan had decreased dramatically during the latter half of the twentieth century, owing to a number of improvements in systems, methodologies, facilities, materials and equipment for the science and engineering of disaster management. These included advanced anti-seismic technologies such as quake-resistant structures and quakeproof retrofitting methods, as well as strict legal regulations on quakeproof construction standards, introduced nationwide in 1981 in response to the serious structural damage that occurred in the 1978 Miyagi-oki Earthquake. Such advances in the scientific control of earthquakes had made many Japanese people, including Kobe residents, less alert, less prepared and more optimistic.

In retrospect, such optimistic attitudes were unfounded; earthquakes posed a risk for all cities in Japan, and always will. Japan is located at the intersection of four interacting tectonic plates (Japanese Cabinet Office 2006), resulting in massive earthquakes due to plate subduction (such as the 1923 Great Kanto Earthquake and the 2011 Great East Japan Earthquake) and inland crustal earthquakes due to plate movements (such as the 1995 Great Hanshin-Awaji Earthquake). Indeed, more than 20% of the world's magnitude 6 or greater earthquakes have occurred in or around Japan.

Earthquakes therefore pose an immediate threat to all major cities in Japan. Based upon the 90–150 year intervals of past earthquake activity and on observations of seismic activity in active faults, earthquake experts predicted that Japan could encounter a major earthquake in the early twenty-first century. Predicted earthquakes include (with the worst affected cities listed in parentheses): a Tokai

Fig. 5.1 Collapse of elevated expressway in the Great Kobe Earthquake in 1995

earthquake (Nagoya and Shizuoka), Tonankai and Nankai earthquakes (Osaka and Kochi), earthquakes around the Japan Trench and Chishima Trench (Sendai) and a Tokyo inland earthquake (Tokyo and Yokohama). Worst case scenarios estimate 20,000–30,000 fatalities and economic losses of more than ¥100 trillion (around US$1 trillion), a figure far exceeding the annual national budget of the Japanese Central Government.

5.1.2 The 2011 Great East Japan Earthquake and Tsunami

On 11 March 2011, 16 years after the Great Hanshin-Awaji Earthquake, a massive earthquake with an epicentre near the Japan Trench, off Japan's northeast coast, caused a tsunami that resulted in unprecedented destruction, particularly along the coast of the Tohoku region. Damage from the tsunami far exceeded that of the earthquake itself; flooding caused approximately 95% of the estimated 20,000 casualties. Economic damage was also extensive, and government estimates are that recovery will cost between 16 and 25 trillion yen (approximately US$200 billion). The Pacific coast of the Tohoku region experiences relatively frequent tsunami events. There have been three significant tsunami since around the start of the twentieth century, in 1896 (approximately 22,000 casualties), 1933 (3,000) and 1960 (150). As a result, the region has implemented both hard damage-prevention measures, such as seawalls, and soft measures, such as evacuation information systems. Most measures, however,

Fig. 5.2 Collapse of a large seawall in Taro district of Miyakoshi City, Iwate Prefecture, after the 2011 Great East Japan Earthquake (Photo by K. Yamori)

were of little benefit in the case of the 2011 tsunami (Fig. 5.2). In addition to damage from flooding, the resulting nuclear accident at the Tokyo Electric Power Company's Fukushima Daiichi Nuclear Power Plant released significant radiation that is likely to extend damage far into the future, both for Japan and the rest of the world.

As discussed above, historical events and ongoing monitoring of seismic activity had suggested that a large earthquake and tsunami would occur in the vicinity of the Japan Trench, with damage centred on Sendai in the Tohoku region. Nevertheless, the scale of the earthquake was surprising. Registering at magnitude 9, it was approximately 45 times more powerful than the 1923 Great Kanto Earthquake, and approximately 1,450 times more powerful than the 1995 Great Hanshin-Awaji Earthquake. The tsunami was over 15 m high in some places, and reached approximately 40 m at its highest point over land. Even specialists did not expect that it would result in a nuclear reactor going into full core meltdown, spreading large amounts of radiation over a wide area (good sources for an overview of the Great Earthquake include *The Japan Times* (2011, 2012) and Kingston (2012)).

Naturally, the Great East Japan Earthquake significantly affected the social representation of earthquake risk amongst the Japanese. Section 5.4 examines this further in consideration of the situation following the Great East Japan Earthquake, after categorisation into four types of earthquake risk images commonly seen in Japanese society.

5.2 The Importance of Sociocultural Views on Earthquakes—From Risk Perception to Social Representation

As Joffe (1999, 2003) observed, mainstream psychological research related to risk has tended to focus upon individual perceptions of risk. Such research assumes the existence of an objective, legitimate risk assessment, as defined by science or specialists with regard to the risk in question, then juxtaposes such an assessment with everyday conceptions of risk held by the general population. These lay perspectives are considered to be distortions of the 'objective', scientific account, caused by problematic subjective biases such as unrealistic optimism (Weinstein 1980). Such frameworks, that conceptualise a legitimate risk assessment (specialised knowledge) standing in contrast to illegitimate risk recognition (everyday knowledge) have been challenged, however, and it is time to move towards research taking the view that both specialised and everyday knowledge concerning risk are historically and socially constructed representations within a given society.

As noted by Moscovici (1976), in pioneering research concerning the social acceptance of psychoanalysis as an example of specialised knowledge, and later emphasised in the framework of social representations theory by researchers such as Jovchelovitch (2002, 2007), Wagner et al. (2000) and Wagner and Hayes (2005), individual risk recognition is a multidimensional and multilayered construct, incorporating knowledge and understanding developed according to criteria different from specialised knowledge. Pertinent examples include folk wisdom concerning methods for tsunami disaster prevention passed on within regional societies, and traditional views of natural disasters as manifestations of fate or divine retribution.

More specifically, research from a social representations theory standpoint has revealed several things. For one, individuals can easily hold specialised knowledge concerning topics involving risk alongside everyday knowledge, even when the two types of knowledge are not necessarily mutually consistent. In some cases, this reflects unique knowledge concerning the topic that has developed historically within the society in question. There are also cases where vernacular science knowledge (Wagner 2007) contributes to unique modifications of specialised knowledge.

Indeed, as noted above, the sudden catastrophe that was the Great East Japan Earthquake had a powerful effect on confidence in specialised knowledge related to earthquakes, tsunami and nuclear power, resulting in a state of clear cognitive polyphasia (i.e. making use of several different modalities of knowledge in approaching a particular issue (Moscovici 1976)) with regard to the image of earthquake risk in Japanese society. This paper avoids relying on the narrowly defined frameworks of risk perception research, or focusing solely upon scientific bases for understanding earthquakes. Instead, it traces traditional views on earthquakes formed over the past centuries, favouring a longitudinal and multilayered approach to investigating the understanding of earthquakes in Japanese society, from the viewpoint of the historical-cultural formation of social representations.

5.3 The Historical Development of the Social Representation of Earthquake Risk in Japan

The historical development of social representations of earthquake risk in Japan can be summarised as occurring in the following four stages: *fatalism* (before the eighteenth century), *social reform* (from the mid-nineteenth century), *scientific control* (after World War II) and *collaborative risk management* (after the 1995 Great Hanshin-Awaji Earthquake). These changes were not straightforward chronological shifts from fatalism to social reform, then onwards through scientific control to collaborative risk management (Fig. 5.3). Rather, these four basic social representations of earthquake risk form a multilayered structure that allows multiple representations to coexist. In this section, we describe each of the social representations of earthquake risk in detail. In the next section, we discuss how the four styles coexist in modern Japanese society, with especial reference to conditions after the 2011 Great East Japan Earthquake.

5.3.1 Fatalism

Old documents and drawings indicate that fatalism was historically dominant in Japan (Hiroi 1986; Kitahara 2000). During the fatalism stage, people were simply scared of earthquakes, and considered prayer as the only means by which they could

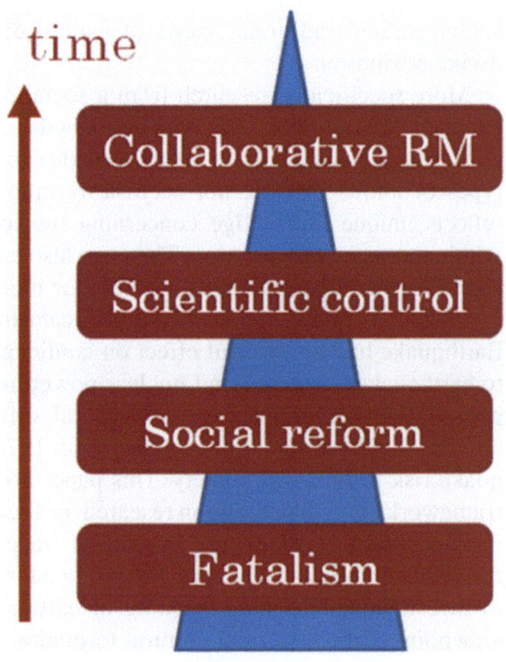

Fig. 5.3 Multilayered structure of social representation of earthquake risk in Japan

Fig. 5.4 A god preparing to shoot a fire arrow as punishment. The image is from a pictorial scroll depicting the Great Kanto Earthquake, painted by Hakudo Kayahara and owned by the Kayahara estate

be avoided. Fatalism related to a view of earthquakes as divine punishment, acts by a wrathful deity from Shinto, traditional local religions, or Confucianism, which was imported from China around the fifth century. Such divine anger was frequently attributed to worldly concerns, such as tyrannical, corrupt, or unstable governance. Such thought can be traced back to the eighth to twelfth centuries, the eras of Nara and the Heian Dynasty (Nakada 1985).

'Divine punishment' was revived with minor modifications in the mid-1920s, following the 1923 Great Kanto Earthquake, which devastated Tokyo. Then, people believed the gods caused earthquakes due to anger about, for example, the lazy, corrupt, or sensuous way people lived in this world. The painting in Fig. 5.4, drawn in the mid-1920s, illustrates this concept. The image shows a god notching a burning arrow to deliver punishment. The flames represent the severe damage caused by the unprecedentedly large-scale conflagrations of the Kanto earthquake, which killed over 100,000 people. Another well-known cartoon (Fig. 5.5) from the same period also shows sentiments of "divine punishment". In this picture, a woman in luxurious clothes is threatened by a god in the form of a catfish. Note that in ancient times, the movements of catfish (*namazu*) living underground were thought to be a cause of earthquakes (see Sect. 5.3.2, for details).

These images suggest that aspects of fatalism persisted in twentieth century Japan, and fatalism has never disappeared entirely, even in the modern era. Nakada (1985) reported statistical data obtained from a 1981 questionnaire survey on disaster

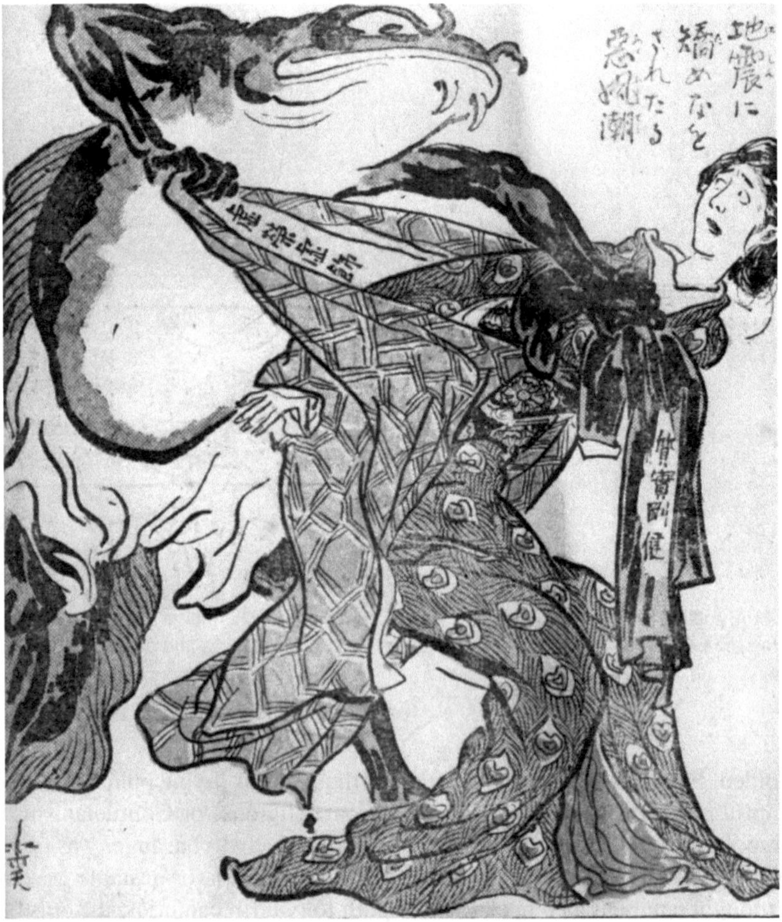

Fig. 5.5 A god in the form of a catfish threatening to punish a lady in luxurious clothes in the aftermath of the Great Kanto Earthquake (Kitazawa 1973, p. 162)

perceptions. One question was, "Do you agree that surviving an earthquake is predetermined as destiny?", to which 35.0% of respondents answered "agree", 29.9% "mostly agree", 18.6% "mostly disagree", 13.4% "disagree", and 3.1% "don't know" or "no answer" ($N=544$). Another question was, "Do you agree that earthquakes are divine punishment?": the percentage of respondents who answered "cannot agree, totally wrong" was 52.4%, with 34.6% for "mostly disagree, but can understand that viewpoint", 8.1% for "mostly agree", 1.6% for "totally agree", and 3.3% for "don't know" or "no answer" ($N=628$).

An Internet questionnaire study ($N=4{,}448$), conducted in 2006 by Tokyo Electric Power Company (2006), found that 52.8% of respondents take some countermeasures against earthquakes, whereas 47.2% do nothing. When those respondents who reported doing nothing were asked in a multiple-response question why they do not take action, 18.6% responded "because earthquake damage is never under human control" (the top reason was "I know something should be done, but do not know exactly what" [50.6%]). These data show that a considerable number of people in Japan maintain a fatalistic mindset even in the present day. Human efforts can reduce earthquake damage, particularly through scientifically-rooted countermeasures (e.g. seismic base isolation), but fatalism persists.

5.3.2 Social Reform

Social reform is a more optimistic and active perspective on earthquakes than fatalism, which often manifests itself as quite pessimistic and passive. As a social representation of earthquake risk, social reform emerged after the Ansei-Edo Earthquake in 1855. This is most clearly symbolised in the ambivalent, dual role catfish play in catfish images (*namazu-e)* (Ouwehand 1964). Japanese historians (see Kitahara 2000) have documented numerous *namazu-e* that were printed and distributed in Edo (now Tokyo) immediately after the 1855 Earthquake. These images provide insight into how the people of that time perceived earthquake risks.

Since catfish were believed to be a major cause of earthquakes, people typically feared them. Simultaneously, however, catfish were seen as harbingers of good – such as the dismissal of an oppressive ruler, a revival of the spirit of mutual aid, or an economic boom – in the aftermath of the earthquakes they were believed to cause (Fig. 5.6). Catfish in *namazu-e* therefore assume an ambivalent, or even contradictory, dual meaning. They are negative, as a destroyer of lives and towns, yet positive, as a trigger for fundamental social reform, freeing people from an oppressed and unhappy world.

As with the fatalistic social representation of earthquake risk, the social reform-oriented representation is not limited to previous eras. It has exercised a lasting influence over the Japanese people since the Edo era, and exists even today. A remarkable twentieth century example was the grand-scale city rebuilding plan that followed the Great Kanto Earthquake of 1923. Shinpei Goto, a well-known politician, and some urban planners considered the earthquake to be a good opportunity to reform old-fashioned Tokyo into a brand-new, modernised capital city. The modern high-rise buildings, advanced subway system and sophisticated underground infrastructure shown in a diorama of the Tokyo Capital Reconstruction Plan (Fig. 5.7) vividly illustrate the optimism people had for the reconstruction of Tokyo after the earthquake.

Just days after the Kanto earthquake, Goto drafted a plan for the reconstruction of Tokyo that included language suggesting a recognition of the close relationship between natural disasters and social reform:

> As befits the capital of the Japanese Empire…the reconstruction of Tokyo shall be no simple matter of restoring a city to its former state, but rather the foundation for expanding the Empire and improving the lives of its citizens. While this earthquake has transformed Tokyo into rubble, and can only be called a great tragedy, it nonetheless provides an ideal opportunity for rebuilding into an ideal capital city. (Koshizawa 2011)

Fig. 5.6 A god riding a catfish scatters Edo-era gold coins. Printed after the 1855 Ansei-Edo Earthquake (Miyata et al. 1995, p. 110)

The scale of Goto's reconstruction plan led to the criticism of it being unrealistic, and consequently some parts were never implemented. Yet, as demonstrated by the fact that 3,200 ha (approximately 90% of the area damaged by fires after the earthquake) were restructured through government action and local cooperation, the plan and the social changes it entailed had popular support (Koshizawa 2011). The plan's legacy persists today, almost a century after its implementation; most of the major traffic routes through modern Tokyo, such as the Showa and Yasukuni roads, were built or expanded as a result of this reconstruction.

The increasingly popular concept of 'pre-disaster reconstruction planning', proposed by disaster management experts (Kato et al. 2008; Nakabayashi et al. 2008) after the 1995 Great Hanshin-Awaji Earthquake, presents a contemporary version of future-oriented social reform expectations. Earthquake reconstruction planning is a kind of pre-disaster planning for post-disaster urban renewal. A recent official reconstruction plan proposed by the Tokyo Metropolitan Government is a typical example (Hosobuchi 2011; Tokyo Metropolitan Government 2003). The plan aims at using the recovery process from projected earthquake damage to solve, or at least improve, serious urban problems such as narrow roads, congested streets, urban

Fig. 5.7 A diorama of the Tokyo capital reconstruction plan (Photo by S. Kondo)

sprawl, antiquated buildings and infrastructure and lack of green space. Such planning is realistic and acceptable, given the high probability of extensive damage from a future Tokyo earthquake.

5.3.3 Scientific Control

Disaster research and management in Japan after World War II was predominantly controlled by central and local governments, because the central government made a political decision to promote disaster prevention work as a major nationwide issue. This administrative policy produced significant achievements. Over time, damage resulting from disasters such as earthquakes decreased dramatically through science- and engineering-centred disaster management (Yamori 2007). For example, during the 25-year period from 1945 to 1969, on average about 1,600 people lost their lives annually in natural disasters in Japan (Japanese Cabinet Office 2010). In contrast, during the period 1970–1994 there were only 215 deaths annually. In 1995 the Great Hanshin-Awaji Earthquake killed more than 6,400 people, but from 1996 to 2010 the annual death toll again nearly halved from that between 1970 and 1994, to 115 people, despite the occurrence of some mid-sized and major earthquakes. This success is primarily attributed to advanced technologies, such as quake-resistant structures and quakeproof retrofitting methods, as well as strict construction standards.

However, such success led to a significant counter-productive result—the reduction of disaster casualties made citizens less alert, less prepared and more optimistic about natural disasters, including earthquakes. During the two to three decades before the 1995 earthquake, people rarely experienced major earthquake damage and came to depend heavily—possibly too heavily—on the scientific control of earthquakes by actors in the public and academic sectors. Palm (1998) reported that the Japanese have a shared general belief that individuals cannot do much to prevent an earthquake from harming them, but that cities, communities and disaster experts can take action to lessen their effects. In reference to Palm's report, Britton (2006) also noted that the Japanese are more inclined than U.S. citizens to let the government take control of disaster-related issues, even if this results in higher taxes.

As mentioned in Sect. 5.3, the fatalistic mindset, so closely interwoven with current social representations of earthquake risk in Japan, has also promoted less alertness and poorer preparedness amongst the Japanese. As scientific control of earthquake damage advances, that which cannot be controlled even by advanced technology gains prominence, strengthening convictions regarding the uncontrollability of natural phenomena by humans.

Another limitation of science-based hard and soft approaches to disaster prevention and mitigation comes from the limitations of the natural sciences themselves. The 2011 Great East Japan Earthquake is a clear example. At a Seismological Society of Japan conference, earthquake researchers publicly called the disaster a "failure of seismology" (Sankei Newspaper 2011), and social criticism of the inadequacy of nuclear power safety controls heightened. An earthquake of magnitude 9 was simply beyond all expectations of researchers, nuclear power plant managers and other specialists, many of whom repeatedly used the word 'unforeseen' (souteigai) in their excuses as to why post-disaster measures were inadequate. The response from the public was that such excuses were simply attempts at avoiding responsibility.

Scientific control undoubtedly remains the predominant social representation regarding natural disasters in modern Japan. The Great Hanshin-Awaji Earthquake and Great East Japan Earthquake, however, have reduced confidence in that viewpoint. This shift was particularly notable after the Great East Japan Earthquake, as demonstrated in an ongoing survey conducted by the National Institute of Science and Technology Policy (2011). When 785 respondents were asked, in November 2010, whether they felt they could trust statements made by scientists, 83.3% said they "could" or "probably could". In April 2011, one month after the disaster, the positive response ratio had fallen to 40.6% (of 756).

5.3.4 Collaborative Risk Management

The 1995 Great Hanshin-Awaji Earthquake resulted in drastic changes. The death toll from this single disaster exceeded the total number of dead in all Japanese natural disasters during the prior quarter century. An elevated expressway, which experts had assured the public was absolutely safe, collapsed (Fig. 5.1). Emergency rescue facilities, such as fire stations and emergency hospitals, to which citizens attached great trust, fell into disarray due to facility malfunctions, personnel shortages and other difficulties. Emergency food and water supplies from outside the disaster area that, according to local government officers, should have arrived rapidly, were delayed due to poor communication and severe traffic congestion. In a well-known book, McCormack (2001) vividly and critically described this paralyzed society as the "emptiness of Japanese affluence".

Although the official response was disappointing, local residents were committed to helping others within their community. More than 70% of those rescued from the debris of collapsed houses were saved by family members or neighbours, not by professional rescue workers (Kawata 1997). Local people also helped each other to survive in emergency shelters without electricity, gas, or sufficient food and water. According to Yatsuzuka (1999), more than 1.5 million people from outside Kobe voluntarily rushed to emergency shelters and temporary housing to help disaster victims. The poor official response raised serious doubts about expert-driven scientific and technological earthquake management systems, with many realising the importance of community assistance after an earthquake.

This resulted in an increased emphasis on citizen-driven, collaborative disaster reduction efforts within local communities. For example, more local governments enhanced cooperation with residents' associations to overcome personnel and resource shortages in local disaster management. Residents also established community disaster management associations and disaster relief non-governmental organisations (NGOs) to better protect their community by themselves (Shaw 2003; Shaw and Goda 2004). The mass media were generally very supportive of these movements.

The sudden rise of voluntarism following the 1995 Kobe earthquake was quite pronounced. As Yatsuzuka (1999) noted, 1995 is often referred to as the year of the "Voluntarism Renaissance" in Japan. After the Kobe Earthquake many people participated in disaster-related voluntary work, and newly established disaster relief-related NGOs organised a nationwide network for mutual assistance in the event of disaster (Atsumi 1997; Atsumi and Suzuki 2003). These NGOs have been quite active and continue to develop, helping disaster victims both in emergency and recovery phases. They contribute particularly towards facilitating local citizens, local government officials, volunteer workers and disaster experts to organise collaborative teams. These NGOs have participated not only in Japanese disasters, but also overseas calamities such as the 2003 Bam Earthquake in Iran, the 2004 Indian Ocean Tsunami and the 2008 Sichuan Earthquake in China.

5.4 The Impact of the Great East Japan Earthquake

The Great East Japan Earthquake had a profound effect on the social representations of earthquake risk in Japan. It is premature to attempt a full description and analysis of its consequences, but a few statements can be safely made. One is that the four predominant social representations of earthquake risk introduced in this chapter continue to coexist, despite the conflicts and contradictions this engenders. In other words, there is a growing trend towards cognitive polyphasia.

Firstly, within the social representations of earthquake risk that have most recently appeared in Japan, one can see both strengthening and weakening trends of collaborative risk management. Unlike the Great Hanshin-Awaji Earthquake, the Great East Japan Earthquake affected a wide geographic area. Damage from the Great Hanshin-Awaji Earthquake was limited to approximately 30 towns and villages. In contrast, even limiting discussion to areas that suffered direct damage from the tsunami following the Great East Japan Earthquake, over 150 towns and villages were affected in that event. This resulted in new types of collaborative risk management. For example, the disaster relief NGOs created after the Great Hanshin-Awaji Earthquake and now experienced in wide-area relief efforts (see Sect. 5.3.3) played an important part in recovery efforts. Google Japan opened its Person Finder system as a means of verifying the status of missing persons; just one example of how the social networking that had rapidly permeated Japan in the decade following the Great Hanshin-Awaji Earthquake provided grassroots tools for disaster response to great effect.

On the other hand, far more on-site volunteerism was seen in the Great Hanshin-Awaji Earthquake than in the Great East Japan Earthquake, despite the scale of the latter being far greater. In the month following the earthquakes, 620,000 volunteers had lent aid for the former, but only 110,000 for the latter. One year after each earthquake, the numbers were 1.37 million and 950,000, respectively (Japan National Council of Social Welfare 2012). Factors such as the extremely large geographical area and worries about the nuclear accident no doubt played a part in these decreased numbers. Still, these numbers indicate that the foundations of post-disaster volunteerism have weakened in Japanese society. One might also suggest that the Great East Japan Earthquake revealed limits to collaborative risk management by local residents; many members of regional volunteer organisations, those whose task it was to aid the elderly in particular, became casualties themselves when the tsunami struck (Yamori, in press).

Sections 5.1.2 and 5.3.3 have already provided some evidence indicating a weakening of reliance upon scientific control, despite its predominance following World War II. To that, this chapter adds two examples that illustrate this phenomenon. The first is the seawall in the Taro district of Miyakoshi City, Iwate Prefecture (Fig. 5.2). At a length of 2.5 km and height of 10 m, the seawall, called 'the Great Wall' locally, surrounded the town, making it something like a walled city. Tsunami had seriously damaged it three times over the past century, leading to its fortification over a period of 50 years. The tsunami that followed the Great East Japan Earthquake destroyed most of it, however, leading to the deaths of almost 200 of the region's 4,400 residents. Some residents even reported that they failed to evacuate the area due to their confidence that such a magnificent wall would surely protect them (Sankei Newspaper 2012); a situation that may have served to consolidate the burgeoning distrust of science- and technology-based hard solutions.

Further lessening such trust was the nuclear accident at the Tokyo Electric Power Company's Fukushima Daiichi plant, and the response to radiation pollution by specialists and the government. Despite previous claims that nuclear power in Japan was perfectly safe, not only did a large-scale accident occur, but furthermore there is no consensus amongst specialists about the safety risks posed by possible future natural disasters. The issue here is not just the large gap between specialists (specialised knowledge) and the understanding of the general populace (everyday knowledge), but the widening gap between the opinions and attitudes amongst holders of specialised knowledge. The former case generally has little effect on fundamental distrust of specialised knowledge (science), but this is not the case for the latter. The current situation is a more serious one that threatens to invalidate the framework that emphasises the importance of risk communication from specialists to the general population with regard to gaps in the understanding of risk. Along with the decreased reliance upon science, and fuelled by increased Internet-based communication, the amount of vernacular science knowledge concerning earthquakes and radioactive pollution has reached unprecedented levels, and blurred the boundaries between pure scientific knowledge and everyday knowledge. For example, alongside the scientifically supported suggestion that iodine tablets can be effective in preventing radioactive contamination, there was a wealth of intuitively plausible but scientifically groundless advice (such as that seaweed consumption was just as effective a measure as an iodine tablet) circulating on the internet and in everyday conversation.

Another important aspect relates to the changes that the Great East Japan Earthquake wrought upon the social reform image of earthquakes, despite its long history. Immediately preceding the quake, Japan ranked third worldwide in the number of operating nuclear reactors—54 plants that together supplied 23% of Japan's electricity. After the earthquake, however, calls for an immediate or staged shutdown and elimination of all nuclear power plants—formerly a radical notion—gained popular support. In a Gallup survey taken after the earthquake (Gallup Inc. 2012), support for nuclear power amongst the Japanese had fallen from 62% in 2007 to just 39%, and those claiming to be opposed increased from 28% before the earthquake to 47% after. On 5 May 2012, all nuclear reactors in Japan were shut down, and due to citizen concerns regarding their safety, it remains unknown when they will be brought back into operation. Something similar was seen immediately following the Great Hanshin-Awaji Earthquake, with calls for radical reform of Japan's transportation system, including the elimination of the Shinkansen rapid rail system and all highway bridges. Whilst most of those plans were never implemented, the post-Great East Japan Earthquake realisation of such vast changes in a society so dependent on high energy consumption, in which nuclear energy played such a vital role, is a clear example of how the earthquake is generating opportunities for movement towards social reform.

Also interesting is the resurfacing of notions of fatalism and divine punishment (see Sect. 5.3.1) that had been lurking in the shadows of scientific control in modern Japan. In centuries past, such notions may have served as a social control mechanism for maintaining social order and feelings of well-being. Humans fear ignorance above all else, and not knowing what something is, or why it occurs, can elicit feelings of a deficit of identity or meaning for that phenomenon. In times when the natural sciences provided no explanation for natural disasters, the sudden disruption of otherwise peaceful lives by such unexplainable events was no doubt troubling. Thus fatalism and divine punishment arise as a defence mechanism against such fears, giving natural disasters meaning along with a possible means of coping; one must submit to inevitable fate rather than fear it, and lead a virtuous life so as not to become the target of punishment oneself. Following the Great East Japan Earthquake, one politician stirred controversy by stating that the disaster was punishment for leading selfish lives, a statement that indicates a mechanism for coping with such an unprecedented situation. The foreign media, as typified by Kingston (2011), made frequent comments on the peaceful and orderly behaviour of the Japanese even in the period of confusion immediately following the earthquake, attributing this to a tradition of fatalism in Japanese society.

Even more interesting modern patterns are the interactions between a sense of fatalism or divine punishment and the reconsideration of excessive, unconditional reliance upon knowledge of disasters provided by the natural sciences. Some felt that the Great East Japan Earthquake was divine retribution for the excessive belief in the supremacy of science and technology that society operates under—that the Japanese had become arrogant in their ability to restrain the power of nature—and this event was punishment for such hubris. The feeling is that no matter what advances science may achieve, humans must live in subservience to nature and can never fully contend with its fury. Terada (2011) and Yamaori (2011) present such views. Somewhat paradoxically, the attribution of natural disasters to divine punishment that characterised earlier times due to a *lack* of scientific understanding now resurfaces in the ascription of such catastrophes to divine punishment for an *overabundance* of, and overreliance upon, this scientific information.

5.5 Concluding Remarks

As discussed above, it is impossible to view the social representation of earthquake risk in modern Japanese society as a simple tiered structure. The four types of representation presented in this paper coexist as a strengthening form of cognitive polyphasia. This has happened because, as noted by Jovchelovitch (2002), humans implement multiple social representations of risk according to the present social context, and some of them may contain logically inconsistent elements. For example, it is not difficult to imagine that, following the Great East Japan Earthquake, a member of Japanese society might invoke scientific control when considering mass media reports of earthquake or tsunami predictions by researchers, collaborative risk management when participating in tsunami evacuation drills alongside other community residents, fatalism when discussing with family members the unavoidability of one's fate when an unexpectedly large earthquake strikes, and social reform when hoping that the nuclear accident will lead to significant improvements in the perceived politically and economically stagnant state of affairs in Japan.

Another factor furthering the trend towards cognitive polyphasia is the increasingly diverse forms of media that affect the formation of images of earthquakes.

The extreme popularity of *namazu-e* in the Edo era (see Sect. 5.3.2) was a result of social permeation of simple layers of fatalism or social reform. The modernisation and advancement of formal education that followed World War II, as well as the influence of mass media such as television and newspapers, helped to rapidly advance notions of the superiority of scientism (scientific control) and voluntarism (collaborative risk management) in simple layers throughout Japan. In contrast, recent advances in social media accessed via portable electronic devices and the Internet have broadened the range of information sources, giving access to varied and sometimes starkly dissimilar or even contradictory scientific knowledge, as well as variant forms of vernacular science knowledge. The coexistence of these two forms of knowledge has helped in forming the cognitive polyphasia seen today.

To grasp the complex structure of risk representation presented in this paper, it is insufficient to consider only immediate, momentary, overt coping behaviour. Instead, it is necessary to examine the evolution of culturally shared representations of earthquakes over a longer time frame. Also necessary is to uncover cognitive polyphasia of both overt and potential earthquake risk images, rather than simply to examine the superficial appearance of risk perceptions or concrete countermeasures against earthquakes. In this regard, the data collection procedure devised by Joffe (2011), eliciting freely associated words and images that are then discussed with the participant, is a promising technique for clarifying such cultural representations of earthquakes.

References

Atsumi T (1997) Disaster relief efforts by Japanese governmental and non-governmental organizations: impact of the Hanshin Earthquake on domestic and international disaster relief. Papers and proceedings of international symposium on disaster and health, Manila, 1997

Atsumi T, Suzuki I (2003) A nationwide network of disaster NPO's to cope with regional vulnerability in Japan. Paper presented at the third DPRI-IIASA international symposium on integrated disaster risk management: coping with regional vulnerability, Kyoto, 2003

Britton N (2006) National planning and response: national systems. In: Rodriguez H, Quarantelli EL, Dynes RR (eds) Handbook of disaster research. Springer, New York, pp 347–367

Gallup Inc. (2012) The impact of The Great Tohoku Earthquake on people's attitude towards nuclear power station. http://donjon.rulez.jp/refeqsum/Gallup-110420.pdf. Accessed 2 Aug 2012

Hiroi O (1986) Disasters and Japanese people. Jiji-Tushin-sha, Tokyo

Hosobuchi J (2011) Pre-disaster recovery planning: Tokyo. Paper presented at the international recovery forum 2011, Kobe, 2011

Japan National Council of Social Welfare (2012) Statistics report on volunteer people in the Great Tohoku Earthquake. http://www.saigaivc.com/. Accessed 2 Aug 2012

Japanese Cabinet Office (2006) Disaster management in Japan. http://www.bousai.go.jp/panf/saigaipanf.pdf. Accessed 2 Aug 2012

Japanese Cabinet Office (2010) White paper on disaster management 2010. http://www.bousai.go.jp/hakusho/h22/index.htm. Accessed 2 Aug 2012

Joffe H (1999) Risk and 'the other'. Cambridge University Press, Cambridge

Joffe H (2003) Risk: from perception to social representation. Br J Soc Psychol 42:55–73

Joffe H (2011) Thematic analysis. In: Harper D, Thompson AR (eds) Qualitative research methods in mental health and psychotherapy. Wiley-Blackwell, Chichester, pp 209–223

Jovchelovitch S (2002) Re-thinking the diversity of knowledge: cognitive polyphasia, belief and representation. Psychologie et Société 5:121–138

Jovchelovitch S (2007) Knowledge in context: representations, community and culture. Routledge, London

Kato T, Nakabayashi I, Ichiko T (2008) Development of planning support system for urban rehabilitation and reconstruction. J Disaster Res 3(6):422–428

Kawata Y (1997) Prediction of loss of human lives due to catastrophic earthquake disaster. J Nat Disaster Sci 16:3–14

Kingston J (ed) (2012) Natural disaster and nuclear crisis in Japan: response and recovery after Japan's 3/11. Routledge, London

Kingston J (2011) Foreign policy: Tsunami–Japan's post-Fukushima future. http://www.foreign-policy.com/files/tutEkfeUr4fOa3v/06282011_Tsunami.pdf. Accessed 2 Aug 2012

Kitahara I (2000) Social history of earthquakes in Japan: the Ansei Edo Earthquake and Japanese people. Kodansha, Tokyo

Kitazawa R (1973) Collection of Rakuten cartoon: Taisho period version. Graphic-sha, Ohamiya

Koshizawa A (2011) Shinpei Goto: the Great Kanto Earthquake and reconstruction of imperial capital Tokyo. Chikuma Shobo Press, Tokyo

McCormack G (2001) The emptiness of Japanese affluence. M.E. Sharp, New York

Miyata N, Kitani M, Imada Y, Takada M, Kitahara I (1995) Catfish cartoon picture: earthquakes and Japanese culture. Ribun Shuppann Press, Tokyo

Moscovici S (1976) La Psychanalyse, Son Image et Son Public, 2nd edn. Press Universitaires de France, Paris

Nakabayashi I, Aiba S, Ichiko T (2008) Pre-disaster restoration measure of preparedness for post-disaster restoration in Tokyo. J Disaster Res 3(6):407–421

Nakada M (1985) Disaster and Japanese people: natural disasters as psychological phenomena. Unpublished manuscript. http://www5b.biglobe.ne.jp/~info-pat/disastersandJ.htm. Accessed 2 Oct 2012

National Institute of Science and Technology Policy (2011) Change of Japanese people's attitude towards science and technology after the 3.11 disaster: based on a monthly attitude survey of science and technology http://www.mext.go.jp/b_menu/shingi/gijyutu/gijyutu0/shiryo/__icsFiles/afieldfile/2011/06/10/1307169_07.pdf. Accessed 2 Oct 2012

Ouwehand C (1964) Namazu-e and their themes: an interpretative approach to some aspects of Japanese folk religion. E.J. Brill, Leiden

Palm R (1998) Urban earthquake hazards: the impacts of culture on perceived risk and response in the USA and Japan. Appl Geogr 18(1):35–46

Sankei Newspaper (2011) "A serious defeat of seismology": unusual self-criticizing statement at the symposium of Seismological Society of Japan. 16 Oct 2011. http://sankei.jp.msn.com/affairs/news/111016/dst11101607570002-n1.htm. Accessed 2 Oct 2012

Sankei Newspaper (2012) Unguarded, unprepared: "No expectation of huge earthquake," "Over-trust on a giant seawall", 10 Mar 2012

Shaw R (2003) The role of non-governmental organizations in earthquake disaster management: an Asian perspective. Reg Dev Dialogue 24(1):117–29

Shaw R, Goda K (2004) From disaster to sustainable civil society: the Kobe experience. Disasters 28(1):16–40

Terada T (2011) Natural disaster and Japanese people. Kadokawa Shoten Press, Tokyo

The Japan Times (2011) The Japan Times special report 3.11: a chronicle of events following the Great East Japan Earthquake. The Japan Times Ltd., June 2011

The Japan Times (2012) The Japan Times special report: 3.11, one year on – a chronicle of Japan's road to recovery. The Japan Times Ltd., April 2012

Tokyo Electric Company (2006) Questionnaire survey on earthquake countermeasures at home: TEPORE members internet research. http://www.tepore.com/research/co/060831/page02.htm. Accessed 2 Oct 2012

Tokyo Metropolitan Government (2003) Preparation for recovery by the TMG: TMG earthquake disaster recovery manual. http://www.bousai.metro.tokyo.jp/japanese/knowledge/material_t.html. Accessed 2 Oct 2012

Wagner W, Duveen G, Verma J, Temel M (2000) I have faith and at the same time I don't believe: cognitive polyphasia and cultural change in India. J Community Appl Soc Psychol 10(4):301–314

Wagner W, Hayes N (2005) Everyday discourse and common sense: the theory of social representation. Palgrave Macmillan, Basingstoke

Wagner W (2007) Vernacular science knowledge: its role in everyday life communication. Public Underst Sci 16(1):7–22

Weinstein ND (1980) Unrealistic optimism about future life events. J Pers Soc Psychol 39(5): 806–820

Yamaori T (2011) Peaceful death in our days. At-pluss 8:154–167

Yamori K (2007) Disaster risk sense in Japan and gaming approach to risk communication. Int J Mass Emerg Disasters 25(2):101–131

Yamori K (in press) Revisiting the concept of tsunami tendenko: Tsunami evacuation behavior in the Great East Japan Earthquake. In: Disaster Prevention Research Institute, Kyoto University (ed) Natural disaster science and mitigation engineering: DPRI reports, vol 1, Investigation on the 2011 off the Pacific Coast of Tohoku Earthquake. Springer, Dordrecht

Yatsuzuka I (1999) The activity of disaster relief volunteers from the viewpoint of social representations: social construction of borantia (volunteer) as a new social reality after the 1995 Great Hanshin Earthquake in Japan. In: Sugiman T, Karasawa M, Liu J, Ward C (eds) Progress in Asian social psychology, vol 2. Kyoyook Kwahaksa, Seoul, pp 275–290

Chapter 6
Facilitating Community Participation in Disaster Risk Management: Risk Perception and Preparedness Behaviours in Turkey

Nuray Karanci

Abstract This chapter aims to present a brief review of disasters and the disaster management system in Turkey, followed by a presentation and evaluation of some psychological models developed for explaining individual disaster preparedness behaviours. The models all stress the importance of facilitating awareness of risks, having information about methods to combat hazard (i.e. what to do) and the role of resources in predicting preparedness behaviours. Also highlighted are the significance of evaluations of coping choices, the perceived efficacy and cost of these choices and the availability of personal (e.g. education; self-efficacy), social (e.g. availability of social networks; civil society organisations dealing with particular hazards) and economic (e.g. financial resources; availability of long term credits) resources. Additionally, some factors that hinder preparedness behaviours, such as helplessness, fatalism, denial and externalisation of responsibility (i.e. belief that mitigation and preparedness is the responsibility of local or central government institutions) are presented. The chapter reports findings from studies on individual and community training and involvement in disaster risk management and predictors of hazard adjustment behaviours in Turkey (mostly conducted in Istanbul). It offers suggestions for community training programmes that aim to facilitate mitigation and preparedness behaviours in individuals and communities.

6.1 Disasters and the Disaster Management System in Turkey

Turkey is prone to various natural hazards and has experienced extensive loss of life and property due to high structural and social vulnerability. Earthquakes, landslides, floods and rock falls are the most frequent types of natural disasters experienced in

---

N. Karanci (✉)
Department of Psychology, Middle East Technical University,
Ankara, Turkey
e-mail: karanci@metu.edu.tr

H. Joffe et al. (eds.), *Cities at Risk: Living with Perils in the 21st Century*, Advances in Natural and Technological Hazards Research 33,
DOI 10.1007/978-94-007-6184-1_6, © Springer Science+Business Media Dordrecht 2013

Turkey (Gülkan 2009). Turkey lies on active fault lines; 96% of the total area of Turkey faces a threat of earthquakes and 98% of the population lives in earthquake-prone areas. Between 1900 and 2010, 253 destructive earthquakes occurred in Turkey, resulting in the death of 94,100 and the injury of 78,808 people (Türkiye Ulusal Afet Arşivi 2012). The greatest devastation occurred as a result of the Marmara and Düzce Earthquakes, which occurred on the 17th August 1999 and 12th November 1999 respectively. The official death toll was 18,000 from these two earthquakes, with a high level of damage inflicted on industrial and communications infrastructure (Türkiye Ulusal Afet Arşivi 2012). Floods are the second most destructive natural disaster in Turkey, with 1,308 floods occurring between 1955 and 2002, killing 1,235 people and causing serious damage to 61,000 dwelling units (JİCA 2004).

Development of the Turkish disaster management system began in 1944 (Gülkan and Karanci 2012). Before 1944, there was no specific structure for disaster management planning and implementation. Disaster management mainly focused on the post-disaster period and consisted of the provision of basic needs (i.e. shelter, food and clothing) after disasters. In 1944 – primarily due to the experience in 1939 of a devastating earthquake in Erzincan, which saw a death toll of 32,962 (Türkiye Ulusal Afet Arşivi 2012) – a specific law was enacted that considered measures to be taken before and after earthquakes (Law No: 4623). This was supplemented in the 1950s by the enactment of the *Development Law* (1956) and the *Civil Defence Law* (1958). In 1959 due to the shortcomings of the 1944 legislation that focused solely on earthquakes and neglected reconstruction, a new law called the *Disaster Law* (Law No: 7269) was enacted. This extended to other kinds of natural disasters and led to the establishment of the Ministry of Reconstruction and Resettlement, which was given responsibility for the law's implementation. Devastating earthquakes in Erzincan (1992) and Dinar (1995) precipitated further development of the laws governing disaster management and in 1997 a new Crisis Management Regulation was put forward as a Cabinet Decree. Following the devastating Marmara earthquake in 1999, the Government issued a number of decrees to deal with the needs of the post-earthquake situation, for example requiring the supervision of building construction practices.

The current disaster management system in Turkey is highly centralised and hierarchical. In order to improve coordination of mitigation, preparedness and response, a new Prime Ministry Disaster and Emergency Management Presidency (Prime Ministry Disaster and Emergency Presidency 2012) was established in December 2009. Although community participation is a necessary component of an effective disaster risk management system, the Turkish Disaster Management system does not contain mechanisms to promote the participation of the communities at risk (Gülkan 2009; Gülkan and Karanci 2012; Karanci and Akşit 2000). The institution of a mandatory insurance scheme in 2000, which is managed by a newly formed Natural Insurance Authority (DASK) (Doğal Afet Sigortaları Kurumu 2012), and requires each urban homeowner to join the Turkish Catastrophe Insurance Pool (TCIP), represented a major move for socially distributing risk and increasing community awareness of earthquake risks (Gülkan 2009). The DASK has also been involved in raising community awareness for earthquakes by using a mobile earthquake simulator, visiting provinces along the northern Anatolian fault-line and giving basic information on earthquakes and preparedness measures. A total of 110,000 individuals were reached through this attempt and were given an experience of quakes in the form of a shaking simulator along with information on preparedness and mitigation. However, the purchase of the insurance is reported to be only 26%. It is highest in the Marmara region (34%), probably due to the 1999 Marmara earthquake (Doğal Afet Sigortaları Kurumu 2012). Furthermore, policies promoting actual community involvement in mitigation and preparedness are still lacking.

Istanbul, which has a population of nearly 12 million and very high inward migration rates (TUIK 2009), has a high probability of experiencing a severe earthquake in the next 30 years (Eraybar et al. 2010; Parsons et al. 2000). Effects of the 1999 Marmara earthquake extended to some parts of Istanbul, and since 1999 there has been an extensive focus on disaster risk mitigation in Istanbul. With a major loan agreement from the World Bank, the Istanbul Seismic Risk Mitigation and Emergency Preparedness Project (ISMEP) was set up to manage the development of emergency management capacity and the structural retrofitting of schools, hospitals, public buildings and heritage sites (ISMEP 2012). The project is implemented and monitored by the Istanbul Project Coordination Unit (IPCU), established under the Istanbul Special Provincial Administration (ISPA). A component of the project that was launched in August 2009 specifically focused on increasing the risk awareness of the public and on the facilitation of community mitigation and preparedness behaviours. Within this framework, handbooks on various aspects of mitigation and preparedness for earthquakes have been prepared and training of community volunteers (labelled as 'safe life volunteers') has been launched. The training, which takes a total of 6 h, covers material on disaster awareness, preparation and mitigation, things to do in the first 72 h after an earthquake, first aid, fire extinguishing and the importance of community involvement in disasters (ISMEP 2012). Unfortunately, the impact of these training efforts has not yet been systematically evaluated. Furthermore, mechanisms for transforming the training of individual safe life volunteers into community groups working together to implement mitigation and preparedness measures have not been developed. To ensure the sustainability of these efforts, it is important to devise mechanisms to facilitate the formation of community groups from these safe life volunteers for mitigation and to monitor and evaluate their outcomes.

6.2 The Role of Community Participation in Disaster Risk Management

For effective disaster risk management to occur, local community members and community groups need to be involved in the necessary mitigation and preparedness behaviours (Basolo et al. 2010; Burningham et al. 2008; Gülkan and Karanci 2012; Karanci and Akşit 2000; Pearce 2003; Solberg et al. 2010). In order to increase community resilience it is vital to facilitate mitigation and preparedness behaviours and collaboration of local communities with disaster risk management agencies. Increasing awareness and facilitating adjustment behaviours of individual community members does not automatically lead to the formation of community groups that will join efforts to prepare particular districts for disasters. Thus, it is important to design programs which encourage individuals to form community groups or join civil initiatives and work for the mitigation/preparedness of their whole communities. Although there have been some efforts in Turkey to support community groups, like the neighbourhood disaster volunteers, the majority of efforts have been targeted at individuals, with the expectation that subsequently individuals will form groups or will join groups involved in mitigation and preparedness. The neighbourhood disaster volunteers program has trained 4,524 volunteers in 95 neighbourhoods, in Kocaeli, Istanbul, Izmir and Yalova provinces. They have also set Neighbourhood Disaster Centres in these 95 neighbourhoods and are continuing to try to institutionalise their efforts (Neighbourhood Disaster Support Project 2012).

Although there have been numerous efforts to facilitate individual/community participation and to increase actual preparedness behaviours of individuals, families and community groups, the results have not been very encouraging (Basolo et al. 2010; Karanci et al. 2005; Peek and Mileti 2002; Rustemli and Karanci 1999). The main attempts at increasing awareness in community members have operated predominantly through *providing information* through various sources, such as training programs and brochures. The evaluations of training programs focusing on facilitating awareness have shown that awareness and information do not automatically lead to preparedness behaviours (Karanci et al. 2005; Paton 2008; Paton et al. 2005). Therefore, it is important to understand the factors that motivate individuals to take action for mitigation and preparedness, as well as factors that hinder such initiatives. Furthermore, we need to understand how community groups operate, what makes individuals join such groups and what facilitates their persistence in their efforts to create a safe environment. Facilitating the adoption of non-structural and structural mitigation and preparedness measures at the individual level at the pre-disaster stage is important to lessen negative consequences of disasters. However, it is also valuable to get individual community members to join community-based organisations (CBOs) focusing on disaster mitigation and preparedness activities, to ensure that whole neighbourhoods are involved.

Most of the psychological models developed to account for engaging in preparedness behaviours explain individual preparedness. The models seem to assume that if individual members of the community are motivated to engage in earthquake protection measures then the negative impacts will be mitigated. However, we need to also focus on CBOs and understand factors that facilitate joining the efforts of such organisations. Inelmen et al. (2004) evaluated factors related to why individuals are reluctant to join CBOs, by examining attitudes towards joining a neighbourhood service cooperative, the Gayrettepe Culture, Environment, and Administrative Cooperative (GC), which is an active local organisation in Istanbul working in earthquake mitigation and preparedness. The main reasons included uncertainty about its goals, unfamiliarity with the GC, lack of time and avoidance. The study also showed that fatalism – a belief that nothing can be done about earthquakes – was prevalent among the individuals living in the neighbourhood and that they expected that most of the pre- and post-disaster measures would be implemented by the state and formal authorities. Thus, they attributed responsibility to these actors. Thus, the mechanisms and incentives for facilitating the involvement of community members in such local civil initiatives needs to be examined further and necessary measures need to be implemented.

While understanding the factors underlying participation at the community level is a critical avenue of investigation, the lack of extant community-level research means that most of the literature reviewed in this chapter relates to factors implicated in individuals' adoption of mitigation and preparedness activities. In the next section some of the models that examine factors related to individual preparedness behaviours will be introduced.

6.2.1 Psychological Models for Preparedness Behaviours

In order to understand factors that facilitate and hinder individual mitigation/preparedness behaviours (also referred to as hazard adjustments), various models have been developed and empirically tested. The Person Relative to Event model (PrE) (Mulilis and Duval 1995, 1997) proposes that risk perception consists of two dimensions: (1) threat and vulnerability appraisals and (2) coping and resource appraisals. Resources can be psychological/personal (e.g. having self-efficacy, optimism, and skills), economic (e.g. financial resources, employment) or social (e.g. social support networks, CBO's working for mitigation/preparedness for disasters). The model asserts that action (i.e. hazard adjustment behaviours) will only take place when coping/resource appraisals are greater than threat/vulnerability appraisals – in other words, when people believe that they have the means to reduce their vulnerabilities. Furthermore, the model clearly highlights the importance of accepting personal responsibility for facilitating action. It has been shown that responsibility for engaging in hazard adjustment behaviours is generally attributed to agents like municipalities, governments and engineers (i.e. externalised responsibility) rather than internalised (i.e. 'I should take precautions myself'), which hinders the individual/family and social groups from engaging in active efforts to lessen the impact of potential hazards and to decrease vulnerabilities (Karanci et al. 2005).

The Protective Action Decision Model (PADM) is a model that was originally developed to explain evacuation decisions and later used for hazard adjustment intentions and actions (Lindell and Perry 1992). According to this model, hazard adjustment depends on two kinds of attributes: firstly, the *hazard-related* attributes such as perceived efficacy of actions to protect people and property (e.g. stabilisation of heavy furniture; not placing heavy furniture in front of doorways, having an earthquake supply bag), and secondly, the *resource-related* attributes such as the cost, skill requirements, time and effort needed for preparedness behaviours (e.g. skill required for stabilisation, cost of purchasing insurance premiums). The model asserts that hazard adjustments are likely to be actualised when hazard-related attributes are evaluated as high and resource-related attributes are evaluated as low or manageable. In other words, preparedness behaviours depend on a cost-benefit analysis. Research based on the model has shown that evaluations of hazard attributes tend to be more important in predicting actual hazard adjustments than resource evaluations (Lindell and Prater 2002). Thus, it is important to increase awareness of the effectiveness of hazard-related attributes rather than simply giving information on what to do. It may be helpful to integrate examples that demonstrate the effectiveness of actions to protect people and property into training programs: for example, it may be valuable to show how insurance helps survivors to compensate for their losses or how stabilisation of furniture prevented injury in past earthquakes.

Protection Motivation Theory (PMT) (Rogers 1983), proposes that in order to facilitate preparedness behaviours, it is crucial to increase the motivation of individuals to protect themselves. The PMT proposes that hazard appraisals and coping appraisals predict protective motivation, which is the motivation of individuals to take action that will protect themselves and their families from future harm. The model also delineates factors that may lessen protective motivation. These are labelled as non-protective/defensive factors, such as denial, fatalism, and unrealistic optimism. Furthermore, resources are moderators for converting protective motivation into protective behaviours. This model has similarities with PrE and PADM in its inclusion of hazard attributes and personal resources, the difference being the distinction between motivation and protective behaviours, and the acknowledgement of the role of hindering factors. Thus, according to the PMT, hazard and coping appraisals lead to protective motivation. For example when a person is aware of an earthquake risk and believes that he/she has the coping capacity to do something (i.e. prepare/mitigate), then the person will have protective motivation to engage in preparedness. However, the person needs to have resources (e.g. money, time, skills) to act on this motivation. Furthermore, the person must not have unrealistic optimism (i.e. 'nothing will happen to me/us'), fatalism (i.e. 'I cannot prevent damage from earthquakes, it is fate that will determine what will happen'), and denial (i.e. 'there is no risk for me/us'). Fatalism – the belief that there is a predetermined destiny and nothing the person does will change this – can become a barrier for mitigation/preparedness, with people with fatalistic attitudes less likely to prepare for earthquakes. In the context of earthquakes, fatalism often entails the belief that all damage is related to earthquake magnitude, which cannot be controlled and thus no preparedness is useful (McClure et al. 2001). McClure et al. (2001) found that fatalism can be reduced by giving high distinctiveness (i.e. only some buildings are damaged), consensus (i.e. certain building designs experience more damage) and consistency (i.e. certain buildings experience damage in different quakes) information, which makes people attribute damage to buildings rather than to earthquakes, which in turn may facilitate mitigation/preparedness.

Finally, the social cognitive model of Disaster Preparedness (DPM) (Paton et al. 2005) represents the process by which preparedness behaviours emerge from factors that may motivate individuals to prepare (e.g. critical awareness, risk perceptions, anxiety), progress into factors for forming intentions to prepare (e.g. outcome expectancy, self-efficacy, action coping) and end in decisions for preparedness. Internalised responsibility, trust in emergency management authorities and a sense of community are factors that link these intentions to actual behaviours. Namely, the model proposes that internalised responsibility and trust in authorities are important moderators between intention and action. The model was largely supported by empirical data collected from households in New Zealand for earthquake preparedness (Paton et al. 2005).

The models presented above all share acknowledgement of the importance of risk or threat perceptions and vulnerability perceptions and appraisals. A second category of factors includes personal resources (such as high outcome and self-efficacy beliefs), knowing about methods of coping, having the skills to cope and having adequate financial and community resources. Personal responsibility for taking action also seems to be an important factor in several of these models.

Although only delineated in some of the above models, research has shown that there are some factors that may hinder preparedness behaviours. These include denial (i.e. not acknowledging the risk), fatalism (i.e. belief that no action can change a predetermined destiny), avoidance (i.e. not thinking or talking about risks), optimistic bias (i.e. a belief that the person is in a better position compared to others and that nothing will happen to them), externalisation of responsibility (i.e. expectation that outside agencies are responsible for mitigation/preparedness measures) and past experience in which no substantial loss was experienced due to a hazard event (Kasapoğlu and Ecevit 2003; Mishra et al. 2009; Rustemli and Karanci 1999; Sattler et al. 2000). It is important that earthquake training programmes also tackle these hindering factors.

6.2.2 *Facilitating Hazard Adjustment Behaviours in Community Members*

The models presented above and empirical research findings suggest that certain elements seem to be important in facilitating community members to adopt preparedness behaviours. Unfortunately, these models do not address community-level outcomes directly. For the individual-level outcomes, firstly, giving information on hazards, vulnerabilities and skills – in other words increasing risk awareness – seems to be vital (Lindell and Prater 2002; Paton et al. 2005). However, this in itself does not guarantee the performance of preparedness behaviours. It is also necessary to increase the awareness of community members with regards to hazard and resource attributes. This can be done by promoting beliefs in the effectiveness of certain mitigation and preparedness measures in reducing damage to life and property. Hence, increasing the awareness of risk needs to be followed by an increase in understanding of what to do and belief in the effectiveness of the measures taken. Other resource variables such as the cost, time and energy needed and the skills necessary to conduct hazard adjustments should also be targeted in training programs. Thus, simple education campaigns that focus on the risks alone are not likely to significantly change behaviour. Especially for earthquakes, local government and other resources may need to be activated to support community members in taking action. For structural mitigation measures local governments can provide technical expertise and low credit loans. Besides all these, responsibility is an important element. Externalised responsibility needs to be fought by finding methods to facilitate community participation. Inelmen et al. (2004) showed that individuals believe that they lack necessary skills to contribute to CBOs. Thus we need to increase skills in community members to motivate them to take part in civil organisations preparing for quakes.

Also requiring consideration are factors that hinder community members from adopting preparedness behaviours such as denial, fatalism and optimistic bias. It is understandable that when individuals believe that they lack resources to engage in mitigation/preparedness behaviours, denial and optimistic bias may serve to reduce their anxieties, providing some respite. In order to be sustainable, measures employed to combat such hindering factors should consider their cultural relevance. For example, in the context of Turkey, in which the majority of the population is Muslim, fatalism can be fought by giving messages that remind people that it is only after they have done everything that they possibly can to protect themselves and their families that God's will can be expected to operate. It seems likely that if this message is given by credible religious authorities, such as the religious leaders (imams), it may be more effective. However, although imams have been targeted as a group of trainers by the Turkish Red Crescent (TRC) for facilitating preparedness in community members for quakes, the impacts have not been yet assessed (Turkish Red Crescent 2012).

6.3 Some Findings from Turkey

The devastating 1999, Marmara (17th August) and Düzce (12th November) earthquakes in Turkey were a turning point for the realisation of the vital importance of community involvement in disaster risk management. The devastating impacts in the five provinces led to the realisation that public institutions cannot possibly answer all the needs in the post-quake period and therefore community efforts are necessary in providing support. Furthermore, the importance of mitigation and preparedness, and thus the pre-disaster period, was recognised. The 1999 earthquakes devastated highly industrialised and densely populated urban areas in Turkey and led to 18,000 deaths and the severe injury of 44,000 people. Many residential and commercial buildings, bridges, motorways and infrastructure were damaged (Bruneau 2002). The earthquake was followed by major attempts at legislative change, structural mitigation and community involvement initiatives. These included an increased involvement of non-governmental organisations in mitigation and preparedness, the formation of neighbourhood volunteer groups, and various community awareness and training programs. Most important was an attempt to foster community involvement through the establishment in August 2009 of the Community Awareness Component of the earlier discussed Istanbul Seismic Risk Mitigation and Emergency Preparedness Project (ISMEP), which was launched with the slogan, "We have joined our hands against the quake for strengthening our future: A step towards safe living". ISMEP contains several components, addressing structural mitigation of buildings, lifelines (e.g. electricity, water supply) and capacity building (ISMEP 2012). Community training (i.e. safe life volunteers) is covered under the capacity building component. Although it is a move in the right direction to focus on residents of Istanbul, the fact that it was launched 10 years after the 1999 earthquakes seems to reflect a slow institutional response. In the following sections, some studies conducted in Turkey will be presented.

6.3.1 Feasibility Study for Retrofitting Selected Residential Buildings in Istanbul (Bakırköy District)

This study was part of a project aimed at evaluating the technical and social feasibility of retrofitting 367 high-risk buildings in Bakırkoy, Istanbul, having secured the consent and consensus of the owners to participate (Johnston et al. 2006; TC Başbakanlık PUB 2005). This was a novel project that attempted to integrate technical feasibility with social feasibility and demonstrated that without the consent, willingness and involvement of the owners of buildings it is not possible to engage in mitigation regardless of the sophistication and detail of the technical solutions.

A total of 4,882 questionnaires were distributed to tenants and owners living in the 367 buildings. Almost half (n=2,429; 49.8%) were returned, of which 69.5% (n=1,689) were owners, and 16.6% (402) were tenants, with the remainder in the "other" category (e.g. not paying rent but living in family-owned flat). The questionnaire contained questions on the socio-demographic characteristics of the sample, their risk perceptions (i.e. awareness of potential for future quakes, how much damage they anticipated), and their views on effective mitigation measures (e.g. retrofitting, demolishing and constructing new buildings). The results of the study showed that the majority of the respondents were long-term urban dwellers in Istanbul and were not planning to move from their buildings (70%) despite having been informed that their buildings were regarded as at high risk in a potential future earthquake. In terms of hazard perception the majority displayed an optimistic bias, and stated that a major earthquake may happen only after 5 years or a longer period. When asked about their vulnerability to a possible future earthquake, however, the respondents gave quite high ratings, indicating that they perceived themselves as vulnerable to future earthquakes. This finding has been replicated in another study in Istanbul (Eraybar et al. 2010). Awareness of vulnerability co-exists with optimistic bias in that the occurrence of a future devastating quake is delayed. This optimistic bias may serve to reduce the likely anxiety felt by the respondents. The respondents showed quite high outcome efficacy (82%), believing that something can be done to mitigate future earthquake damage. This finding seems to point out that they are not fatalistic in their approach to earthquakes and believe that some measures can be taken. However, their sense of self-efficacy (i.e. the belief in their own ability to do something to mitigate damage) was much lower (45%). This finding suggests that although the sample believed in the *possibility* of mitigation they did not evaluate *themselves* as capable of taking action or as having the technical knowledge and financial capacity through which this could be done, which may point to expectations of external responsibility for reducing their vulnerabilities.

Another set of questions related to perceptions of retrofitting. The responses showed that the sample evaluated strengthening the current building (30%) and/or demolishing the current building and constructing a new building (27%) as effective solutions for mitigation. Of those who evaluated retrofitting as effective, 43% stated that it would be effective for risk reduction and 10% stated that it would be effective for providing psychological comfort. However, they also pointed out the difficulties of cost (35%), having to temporarily move out from their homes (20%), and technical, legal and trust problems (i.e. trust in construction firms and municipality) in the implementation of retrofitting. Most would not want to move out of their buildings during retrofitting and stated that retrofitting should be done while they stayed in their homes (75%).

In the sample as a whole, a majority voiced concerns related to the costs (93%) and wanted suitable credit support (92%) to go ahead with implementing the procedure; while people believed that retrofitting would be an effective mitigation measure, most did not believe that it would increase the value of their buildings (65%). Furthermore, trusting the firm which develops and implements the retrofitting project and trust in the firm supervising the process (90%) appeared to be very important concerns. It is worth noting that counter to some results in this study, Eraybar et al. (2010) found that seismic strengthening was the preference of only a small minority of respondents (8%). It is possible that the favourable attitudes to retrofitting in the present study may have been inflated by asking about the value of retrofitting directly without giving respondents a choice to indicate which among several methods of mitigating earthquake damage they prefer.

These results support the importance of evaluating the *social* feasibility of mitigation measures. Although the owners and tenants living in high risk buildings in Bakırköy, Istanbul were aware of the hazards and their vulnerabilities and believed that mitigation was possible, they had low self-efficacy (i.e. the belief that they could do something about the situation) and required the fulfilment of several conditions before going ahead with retrofitting. Thus, a purely technical analysis, and suggestions based on this, may not be adequate to ensure support for action (i.e. retrofitting). At the end of the project the owners were provided with a detailed analysis of their buildings and the required retrofitting proposals for their buildings. However, due to the lack of suitable credit support for the retrofitting project, it could not be carried out.

Three years after the original data collection, a small sample of residents from these buildings were interviewed about their risk perceptions and their engagement in any hazard adjustments (Harma et al. 2009). From these in-depth interviews it was clear that little mitigation activity had been embarked upon during the intervening 3 years. Data showed that there was partial avoidance of risk awareness, as the following quote illustrates.

> We do not talk about the earthquake since we do not want to harm our child's and others' psychology. But, we still bear it in our minds.

While risk perception seemed to be high, optimistic bias and fatalism were present among the residents.

> When the earthquake occurs, it will be really terrible. Nothing will remain standing, all hospitals, bridges, roads will be collapsed. I do not want to imagine them. One side of this apartment is flat but the other side is overhang. I think this building will fall down in such a way that perhaps, we will not be affected. But, the upper floor, I think will fall down in this direction. I am begging to my God.
>
> If I ask myself whether I do something to cope with the earthquake, I can say that I generally pray to God for protecting us.

There seems to be anxiety about future earthquakes. However, beliefs about the efficacy of retrofitting do not seem to be very favourable. Self-efficacy about conducting mitigation seems to be low, as found in 2005, and points to the importance of cost and thus the necessity of financial support for hazard adjustments.

> I am often anxious about the earthquake. Moreover, I use medicines to cope with this anxiety.
>
> I think that retrofitting is like a ragged suit. And ragged suit cannot be the same as a new one, no matter what you do!
>
> I have only 600 liras in my pocket. So, I cannot think about my life, I consider only my stomach. I cannot think of anything else!
>
> I accept fate and I am just staying.

A last point that emerged from the qualitative analysis is that responsibility was externalised for mitigation and there were problems with trust.

> I think that municipalities should do something. The citizens cannot do anything. There are financial problems.
>
> I don't trust anyone. Even the municipality could aim to profit by directing us to their own firms.

Thus, the original study and the follow-up qualitative analysis supported the importance of not only risk awareness (which is already quite high) but also lack of financial resources, externalised responsibility and lack of trust as hindering factors for adopting hazard adjustments. Furthermore, denial of vulnerability may be an additional hindering factor. Eraybar et al. (2010) found that 80% of their respondents believed that their buildings were strong enough to survive quakes. Thus, programmes aiming to facilitate community involvement need to address all these points and be supported by municipalities in terms of financial resources and technical assistance. Structural mitigation measures can exceed the financial capacities of families and thus low credit, long-term loans may provide them with the financial resources to engage in mitigation.

6.3.2 Other Findings from Turkey

Okman-Fişek and Kabasakal (2008) conducted a comprehensive study in Istanbul assessing various aspects of risk assessment and preparedness behaviours. They identified high levels of risk perception and a strong belief about the importance of safe buildings as a mitigation measure for future earthquakes. However, the respondents estimated that the risk for their own buildings was lower than their general risk estimate for Istanbul, indicating that although there is awareness of risk there is also optimistic bias for their own situation. With regards to responsibility for mitigation, the majority of respondents believed that it rests outside of themselves in others (such as the municipality and government) and only a minority believed that it was their own responsibility. Although risk perception was quite high, few engaged in mitigation measures. Only 20% stated that they had taken action for preparedness/mitigation. The most common reasons for not preparing were "We trust our building" (54%), "It is too expensive" (54%), "It is God's will, what I do won't matter" (41%), "It is of no use" (33%) and "None of my relatives/friends did something" (30%). These responses show that although general risk perception is high, when it comes to individually-faced risks, optimistic bias, fatalism and financial constraints appear as hindering factors. Therefore, community education programmes need to combat unrealistic optimism and fatalism, and give clear information on the effectiveness of mitigation/preparedness measures. Furthermore, attractive financial support in the form of low credit loans needs to support these programs.

Tekeli-Yeşil et al. (2010) also examined factors affecting the process of taking action at an individual level for mitigation and preparedness in Istanbul. The qualitative results from a sample from two districts of Istanbul (Bakirkoy and Beykoz) revealed three patterns that related awareness to mitigation and preparedness actions. The most common pattern was an awareness of the risks related to earthquakes not being followed by any action. The second pattern was awareness followed by some preliminary steps in taking measures but then followed by no action. Finally, the least common pattern was the completion of the sequence from awareness to action. The study revealed that lack of outcome expectancy (i.e. absence of belief in the effectiveness of micro-scale, individual and non-structural measures), helplessness and a related fatalism (i.e. feeling helpless about what to do about earthquakes and then adopting a fatalistic coping style), low socio-economic level, culture of negligence in the context of multiple risks in the community, lack of trust in agencies and institutions that plan and implement retrofitting and other preparedness programmes, and normalisation bias were factors that operated in the first two patterns (i.e. no action). On the other hand, for the third pattern, in which awareness translates into mitigation/preparedness actions, facilitating factors were: direct personal experience with earthquakes, higher education, moderate to high socio-economic level, social interactions involving earthquake mitigation and preparedness discussions and location of house in a high-risk area. Some of the results from this study corroborate the results of the Bakırköy feasibility study (Johnston et al. 2006; see Sect. 6.3.1), in that cost of retrofitting and trust in construction firms and the municipality also appeared as important barriers (indicating that if cheap retrofitting credit were provided, people would retrofit). The importance of educational level, employment, responsibility and awareness as predictors of earthquake preparedness has also been identified in other studies in Turkey (Karanci et al. 2005; Kasapoglu and Ecevit 2003; Rustemli and Karanci 1999).

Taken together, the findings of these studies indicate that although there is an awareness of earthquake risks in Istanbul and in other provinces of Turkey (Gungormus et al. 2012), the level of community members' engagement in preparedness behaviours is low. In other words, awareness does not necessarily and easily convert into earthquake adjustment behaviours. One important issue is whether natural disasters, especially earthquakes, are an important concern for the community members. Eraybar et al. (2010), in their study on risk perception and mitigation behaviours of Istanbul residents, found that earthquakes were perceived as the most significant category of hazard event (46%), followed by disease (31.7%). Thus, potential earthquakes are important for individuals living in Istanbul. Low levels of preparedness and mitigation behaviours therefore do not seem to relate to low prioritisation of earthquakes, but rather to limited financial resources and lack of knowledge how to engage in mitigation, externalized responsibility, low trust in municipality and construction firms, denial and optimistic bias and lack of organisational support from local governments. These factors would need to be addressed in the future in order to facilitate preparedness/mitigation behaviours.

6.4 Conclusions: Barriers and Facilitators of Active Community Involvement

Models on the predictors of individual preparedness/mitigation behaviours point to the importance of awareness as well as having coping resources in facilitating hazard adjustments. From the studies covered in the previous section we can conclude that awareness is sufficiently high – that is, individuals living in Istanbul seem to be aware of their earthquake risk and the devastation it may cause. However, even in this aspect there are some hindering factors, like optimistic bias, which leads individuals to postpone the expected date of a future devastating earthquake. Furthermore, the optimistic bias is also reflected in their lower risk ratings for their own buildings in comparison to other buildings. Low self-efficacy of the residents also seems to be a likely barrier for the implementation of adjustive behaviours. They believe that something can be done, but that this will exceed their financial and personal resources (e.g. technical skills, support networks). They estimate that the cost will be too great. Thus, financial incentives need to be provided and information on actual costs needs to be given. Another component of low self-efficacy may be related to lack of skills in conducting necessary mitigation measures, which can be changed with practical training programs. People may be especially ready to take action for non-structural mitigation measures, such as securing heavy furniture. Thus, giving applied training on how to secure furniture may increase self-efficacy beliefs.

Fatalism and the externalisation of responsibility also seem to be hindering factors. The respondents in all the research studies seem to expect the government or the municipality to take the lead in implementing mitigation or finding suitable credit support. As raised by Tekeli-Yeşil et al. (2010), "a culture of negligence" seems to be present in community members. There are too many hazards (e.g. disease) and difficulties which may be overwhelming for community members. From this a feeling of helplessness or fatalism ("if it will happen it will happen") may develop. Alternatively, such a situation may engender denial of risk ("nothing will happen"). In this context, denial and/or fatalism can serve as emotion-focused coping strategies that may reduce anxiety and thus soothe individuals. Therefore, in all community training programs it is important to combat anxiety, negligent attitudes and fatalism. The involvement of religious leaders in mitigation/preparedness programs may make the messages about the necessity of mitigation taken by individuals more credible and may help to combat fatalism. Especially in the context of the Islamic religion, the Qur'an (holy book of Islam) specifies that first the individual must do everything that they possibly can do to deal with hazards, and then leave it to God's will. Thus, the message given by religious leaders may motivate individuals to take action.

We have learned through our experience of the 1999 earthquakes in Turkey that mitigation/preparedness is crucial for disaster risk management and without the involvement of community members it is almost impossible to effectively manage potential risks related to natural hazards. Thus, a move towards facilitating community involvement is highly desirable. This move needs to be supported by changes in legal frameworks, coupled with political determination and will. The inclusion of community members and organisations by empowering them needs a sustainable effort that is embedded in an institutional framework. The task is an enormously difficult and challenging one, but its products will undoubtedly be worthwhile.

References

Basolo V, Steinberg LJ, Burby RJ, Levine J, Cruz AM, Huang C (2010) The effects of confidence in government and information on perceived and actual preparedness for disasters. Environ Behav 41(3):338–364

Bruneau M (2002) Building damage from the Marmara, Turkey earthquake of August 17. J Seismol 6(3):357–377

Burningham KJ, Fielding J, Thrush D (2008) It'll never happen to me: understanding public awareness of local flood risk. Disasters 32(2):216–238

Doğal Afet Sigortaları Kurumu (2012) Natural Insurance Authority. http://www.dask.gov.tr/. Accessed 10 May 2012

Eraybar K, Okazaki K, Ilki A (2010) An exploratory study on perceptions of seismic risk and mitigation in two districts of Istanbul. Disasters 34(1):71–92

Gülkan P (2009) Disaster management and mitigation systems: a (limited) global picture. In: Gülkan P, Başbuğ Erkan B (eds) Perspectives in disaster management. METU Press, Ankara

Gülkan P, Karanci AN (2012) Grassroots participation vs. dictated partnership: anatomy of the Turkish risk management reality. In: Haque CE, Etkin D (eds) Disaster risk and vulnerability: mitigation through mobilizing communities and partnerships. McGill University Press, Montreal, pp 137–153

Gungormus Z, Karabulutlu EY, Yıldız E (2012) Determining the knowledge and behavior of the individuals about earthquake preparedness at home in Turkey. HealthMED 6(1):232–237

Harma M, Salman S, Aktan T, Elgin VM Karanci NA (2009) Deprem Riski Algısı Zarar Azaltma Davranışlarını Garanti Eder Mi? (Does earthquake risk perception guarantee mitigation behaviours?). Paper presented at the 11th round table meeting, DMC, METU, Ankara, 2009

significantly correlated with each other and also with adoption intention and actual adjustment. The resource-related attributes were also significantly correlated with each other, but were not significantly correlated with adoption intention or actual adjustment. Lindell and Prater (2002) replicated Lindell and Whitney's (2000) findings in a study of households exposed to earthquake hazard in three Southern California cities and three Western Washington cities. This study also found that all three hazard-related attributes had large and significant correlations with each other. Moreover, all three hazard-related attributes had large and statistically significant positive correlations with adoption intention. They also had statistically significant correlations with actual adjustment, but the correlations were much smaller than with adoption intention. However, Weinstein et al. (2000) reported that two measures of perceived control—damage control and injury control—had nonsignificant correlations with "doing something" about tornado hazard.

Research on resource-related attributes (cost, knowledge and skill requirements, time requirements, effort requirements, and required cooperation with others) has reported that all resource-related attributes had large and significant correlations with each other but non-significant correlations with adoption intention or actual adjustment (Lindell and Prater 2002; Lindell and Whitney 2000; Terpstra and Lindell, in press). This is surprising because it is contrary to the PADM's theoretical predictions (not to mention the fundamental assumption of the entire field of economics). More recently, Terpstra and Lindell (in press) replicated these findings of nonsignificant correlations of resource-related attributes with hazard adjustment adoption in a study of a different hazard (flood rather than earthquake) in a different country (the Netherlands rather than the United States). Blanchard-Boehm et al. (2001) found that the cost of past hazard adjustments was not significantly correlated with insurance purchase.

In further analyses of the data previously reported by Lindell and Prater (2002), Lindell et al. (2009) showed the three hazard-related and four resource-related attributes strongly differentiated among the hazard adjustments, as indicated by significant differences among hazard adjustments in respondents' mean ratings on each attribute. These results suggest that both the hazard-related and resource-related attributes were meaningful to the respondents in comparing the hazard adjustments to each other. Moreover, the reanalyses explained previous reports of low correlations between the resource-related attributes and adjustment adoption (Lindell and Prater 2002; Lindell and Whitney 2000). The variation in the ratings of the resource-related attributes (an average of 38.8% of the scale range) is much smaller than the variation in the ratings of the hazard-related attributes (an average of 65.0% of the scale range), which is consistent with the well-known principle that variance restriction can attenuate correlations (Cohen et al. 2003; Nunnally and Bernstein 1994).

In addition, the complexity of the individual hazard adjustment profiles provided a partial explanation for the failure to replicate the three distinct types of hazard adjustments (basic survival, planning and hazard mitigation) reported by Russell et al. (1995). A set of hazard adjustments will load on a single factor only when the attributes on which they are rated are highly intercorrelated with each other. Such a

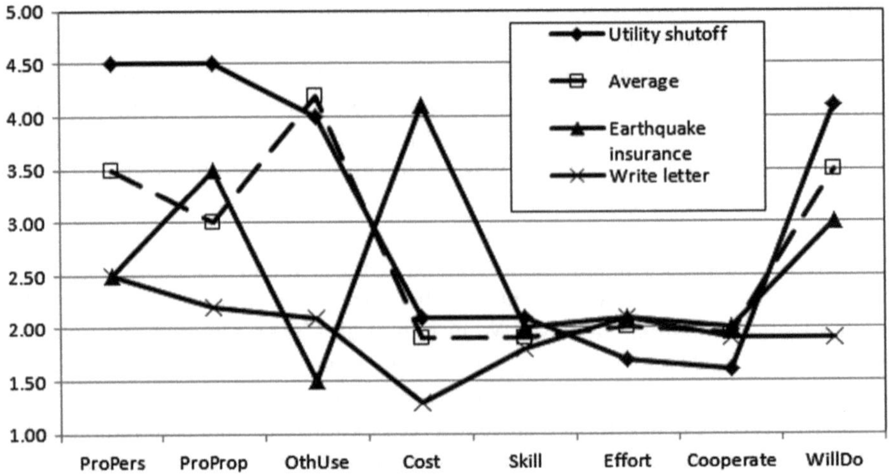

Fig. 7.2 Mean ratings of three hazard adjustments in comparison to the average ratings for all eight

pattern is most likely when all the hazard adjustments in a set have similar profiles across the attributes. However, many adjustments had quite distinct profiles. For example, Fig. 7.2 shows the mean ratings of three earthquake hazard adjustments (knowing where to shut off utilities such as water, gas, and electric power; purchasing earthquake insurance, and writing a letter supporting seismic hazard reduction) in comparison to the mean ratings for a set of eight seismic hazard adjustments (Lindell and Prater 2002; Lindell et al. 2009). These hazard adjustments were rated on their effectiveness in protecting persons (ProPers), effectiveness in protecting property (ProProp), utility for other uses (OthUse), cost (Cost), knowledge and skill requirements (Skill), time and effort requirements (Effort), required cooperation with others (Cooperate), and likelihood of adoption (WillDo). As Fig. 7.2 indicates, there were notable differences among the hazard adjustments in their mean (across respondents) ratings on each of the attributes. Moreover, the profiles show a striking degree of nonparallelism, indicating that the respondents had a complex set of beliefs about the attributes of those hazard adjustments.

7.2.3 *Stakeholder Characteristics*

Research on environmental hazards has long recognized that the adoption of hazard adjustments takes place in a social context involving many different stakeholders, which Drabek (1986) categorized as civil authorities, news media, and peers. There has often been a presumption that authorities are more "credible" than the news media and peers, but hazards researchers have conducted few studies on perceived stakeholder characteristics. The interrelationships among stakeholders can be understood in terms of the Godschalk et al. (1994) "onion theory", in which households (self and family) are located in the central ring, peers (friends, relatives, neighbours, and co-workers) are in the secondary ring, news media are in the tertiary ring, and authorities are in the outermost ring (Arlikatti et al. 2007). The interrelationships among stakeholders can be defined by the power they exert over each other's actions. Although Godschalk and his colleagues did not specify the nature of these power relations, French and Raven (1959, Raven 1965) provide some insights. Specifically, they posited that power relationships can be defined in terms of six types of power—reward, coercive, legitimate, referent, expert, and information power. Extrinsic rewards and punishment (i.e. coercive power) are the principal bases of regulatory approaches, but Raven (1993) noted that these require continuous surveillance to ensure that rewards are received only for compliance and that punishment will follow noncompliance. Most household hazard adjustments are adopted voluntarily in the anticipation of the intrinsic reward of reduced risk. This multiplicity of power bases indicates a need to better understand the ways in which households can be influenced by bases of power other than reward (economic incentives) and coercion (regulations).

French and Raven's conception of expert (i.e. understanding of cause and effect relationships in the environment) and information (i.e. knowledge about the current state of the environment) power suggests assessing perceptions of stakeholders' seismic hazard knowledge. One study examined perceptions of stakeholder hazard knowledge by risk area residents near Mt. St. Helens 5 years after the major eruption (Lindell and Perry 1992). Risk area residents thought there were large differences among federal, state, and local government, the news media, peers, and family/self in knowledge about unfamiliar hazards such as a radiological accident at the nearby nuclear power plant or chemical accident on a nearby interstate highway or rail line. However, the differences in hazard knowledge were perceived to be much smaller for the more familiar hazard of a volcanic eruption—presumably because volcanic activity had stimulated information dissemination by authorities and information seeking by local residents that narrowed any pre-eruption differences in hazard knowledge.

French and Raven's conception of referent power is defined by a person's sense of shared identity with another (Eagly and Chaiken 1993), which is related to trust in that person—especially in their willingness to communicate information accurately. Some insights into stakeholder trust can be drawn from Perry and Lindell's (1986) examination of local residents' perceptions of which was the most credible (i.e. "trustworthy and reliable") source of information about Mt. St. Helens (County Emergency Services, County Sheriff, US Geological Survey, Army Corps of Engineers, friends/neighbours, and TV/radio) and the reason why that source was regarded as credible (special skills/information, concern for citizens, past reliability, and integrity). Over 66% of the nominations for most credible went to the County Sheriff and 63% of the respondents cited elements of trustworthiness (concern for citizens, past reliability, and integrity).

Finally, French and Raven (1959) defined legitimate power in terms of the rights and responsibilities associated with each role in a social network, which raises questions about what households consider to be the responsibility of different stakeholders for protecting them from seismic hazard. As noted by Lindell and Perry (2000), this is reinforced by research on stakeholders' perceived *protection responsibility*, which dates from Jackson's (1977, 1981) research that attributed low rates of seismic adjustment adoption to respondents' beliefs that the federal government was the stakeholder most responsible for coping with earthquakes. Later, Garcia (1989) found that a large percentage of respondents had come to believe disaster preparedness was an individual's responsibility.

More recent research on seismic and other hazards has emphasized perceptions of protection responsibility, although there are some findings on expertise and trustworthiness as well. In particular, the conclusion that perceptions of personal protection responsibility lead to a higher level of seismic adjustment adoption is supported by findings on tornado adjustment adoption (Mulilis and Duval 1997; Mulilis et al. 2003). Moreover, Perry and Lindell (2008) reported significant positive correlations of personal protection responsibility with multi-item scales for wildfire, earthquake and volcano preparedness. Basolo et al. (2009) examined the relationship of perceptions of government protection responsibility with four types of household hazard adjustments—development of a family emergency plan, acquisition of 11 emergency response supplies, adoption of three mitigation measures, and knowledge of how to shut off utilities—for earthquake and hurricane hazards. However, they found a statistically significant positive effect only for development of family hurricane plans. Moreover, Laska (1990) reported a nonsignificant correlation of personal protection responsibility with a flood hazard coping scale.

Other recent studies have examined multiple stakeholder characteristics, hazard knowledge, trustworthiness, and protection responsibility. Lindell and Whitney (2000) reported that hazard adjustment adoption was significantly related to family and personal earthquake knowledge, and personal protection responsibility. Similarly, Arlikatti et al. (2007) found that respondents' ratings of local officials', employers', peers', and family/self's seismic knowledge were correlated with hazard adjustment adoption, as were employer, peer, and family trust, and employer, peer, and family/self protection responsibility. Lindell and Perry (1992) characterised risk area residents' bases for judging risk communicators' credibility as credentials, the respect given them by other information sources, and past history of job performance. This typology is similar to Kramer's (1999) typology of bases for trust within organizations as dispositional, history-based ("past history of job performance"), third party-referenced ("treatment by other information sources"), categorical, role-based ("credentials"), or rule-based.

Arlikatti et al. (2007) also examined whether perceived stakeholder characteristics are directly or indirectly related to households' adoption of seismic hazard adjustments. A direct effect occurs if a stakeholder characteristic influences hazard adjustment adoption through what is described by Fishbein and Ajzen (1975) as the effect of the subjective norm, by Petty and Cacioppo (1986) as the peripheral route to persuasion, and by Chaiken (1987) as heuristic processing. An indirect effect occurs if perceived stakeholder characteristics change people's risk perceptions and this change, in turn, affects their adoption of hazard adjustments. Arlikatti et al. (2007) found that perceived stakeholder characteristics can affect hazard adjustment adoption both directly (Petty and Cacioppo (1986) peripheral route,) and indirectly (via the central route). This finding, which is similar to Gladwin et al. (2001) results for responses to hurricane evacuation warnings, poses an interesting challenge because it means researchers must identify the conditions under which direct or indirect effects should be expected as well as the bases from which the stakeholder characteristics themselves are inferred.

7.3 Antecedents of the Perceptions of the Core Elements

7.3.1 Hazard Intrusiveness

Perry and Lindell (1990) reported that hazard salience—measured in terms of whether respondents thought about volcano threat rarely, periodically, or often—was significantly related to hazard adjustment adoption. Later, Lindell (1994) found that frequency of thought and discussion about a hazard was related to the expectation of severe personal consequences from a hazard. Based on this work, Lindell and Prater (2000) proposed that the frequency of thought, discussion, and passive information receipt about a hazard serves as a reminder for people to take actions that will reduce their hazard vulnerability. Their data showed that this construct, which they identified as hazard intrusiveness, was indeed significantly positively correlated with seismic hazard adjustment adoption. At about the same time, Weinstein et al. (2000) independently found that preoccupation—defined by frequent thoughts, vigilance, intrusive thoughts, and talk frequency—was significantly related to "doing anything" about hazard vulnerability after being struck by a tornado. Similarly, Blanchard-Boehm and Cook (2004) reported that recent information was also significantly related to "doing anything" about hazard vulnerability after being struck by a tornado.

7.3.2 Hazard Experience

Hazard experience can be defined by the recency and frequency of casualties and damage experienced by the respondent him/herself, by members of the immediate or extended family, or by friends, neighbours, or co-workers. Studies have defined earthquake experience using several indicators, such as the number of earthquakes experienced (Russell et al. 1995), amount of previous earthquake losses (Jackson 1981), and experience of earthquake losses by one's self or close others (Turner et al. 1986). Despite the use of varying operationalisations, most studies have found that earthquake experience is significantly correlated with the adoption of seismic hazard adjustments (Lindell and Perry 2000). This conclusion is supported by more recent research on earthquakes and other environmental hazards. For example, Blanchard-Boehm (1998) reported that past damage was significantly correlated with having "done anything" about earthquake hazard and Blanchard-Boehm et al. (2001) reported that past damage was significantly correlated with flood insurance purchase. Sattler et al. (2000) reported a significant correlation of past hurricane damage with preparation of emergency supplies. Nguyen et al. (2006) reported that post-earthquake preparedness measures were significantly correlated with shaking intensity, physical injury, emotional injury, and financial injury. Lindell and Prater (2000) found that earthquake preparedness was correlated with property damage or injury to self, family, or peers. Heller et al. (2005) reported significant correlations of past damage with both pre- and post-impact seismic hazard adjustments and Mullis et al. (2003) reported that having tornado experience was correlated with preparedness scores. Finally, Perry and Lindell (1990) reported that past damage was correlated with the number of volcano preparedness actions taken.

Some studies have found mixed results, with Blanchard-Boehm and Cook (2004) reporting that past damage (but not being in the path or seeing the tornado) was significantly correlated with having "done anything" about tornado hazard. Laska (1990) reported that scores on her coping index were positively correlated with the severity and damage cost of past floods but not their frequency. Norris et al. (1999) found significant correlations of past disaster experience (injury, damage, loss of sentimental items) with hazard alertness and maintenance of basic supplies but not with advance planning. Peacock (2003) reported a significant correlation of hurricane experience with shutter installation but not envelope protection (reinforcing roofs and walls) and Faupel et al. (1992) reported a significant correlation of past experience with preparation of basic supplies but not household planning. Lindell and Hwang (2008) reported significant correlations of flood and wind injury and damage to self and peers with flood and wind adjustments but only flood damage predicted flood insurance purchase. Weinstein et al. (2000) found that "doing anything" about tornado hazard was significantly correlated with having recent reminders, engaging in volunteer acts following previous tornadoes and having contact with victims, but not with having watched the impact or experienced problems later. Perry and Lindell (2008) reported that earthquake preparedness was significantly related to both health and property impacts, and wildfire and volcano preparedness were significantly related to property but not health impacts. Siegel et al. (2003) reported that emotional injury but not physical injury or damage in two previous disasters predicted preparedness.

Finally, there was one study with exclusively negative results (i.e. only nonsignificant correlations). Basolo et al. (2009) found nonsignificant correlations of hazard experience with all four indices—family plan, basic supplies, mitigation, and shutting off utilities.

7.3.3 *Hazard Proximity*

There is mixed evidence regarding the correlations of hazard proximity with hazard adjustment. Farley et al. (1993) reported that adoption of adjustments was correlated with proximity to the New Madrid earthquake fault and Preston et al. (1983) found effects of proximity on perceived personal risk for floods. Similarly, Lindell and Hwang (2008) found that proximity to inland flood and coastal hurricane hazards was significantly related to flood insurance purchase and to the adoption of both flood and hurricane adjustments. By contrast, Palm et al. (1990) and Mileti and Darlington (1997) found no association with proximity to an earthquake fault. More recently, Peacock (2003) found that both hurricane shutter installation and envelope protection were significantly related to location in a coastal county but not location in an evacuation zone.

7.3.4 *Summary of Research on Core Perceptions and Their Antecedents*

The results of research that has correlated the core perceptions (perceptions of the risk, hazard adjustment attributes, and stakeholder characteristics) and their presumed antecedents (hazard intrusiveness, hazard experience, and hazard proximity) with hazard adjustment adoption are summarised in Table 7.1. As the table indicates, hazard adjustment adoption is significantly associated with risk perceptions (65% of all studies in which it was measured), adjustment perceptions (63%), and stakeholder perceptions (67%)—although adjustment perceptions and stakeholder perceptions were measured in only a few studies. Hazard adjustment adoption is also significantly associated with hazard intrusiveness (100% of all studies in which it was measured), hazard experience (48%), and hazard proximity (57%) when these variables are measured—although hazard intrusiveness has been measured too infrequently to draw firm conclusions.

7.4 Demographic Characteristics

There are decidedly mixed findings on the relationships of demographic variables with hazard adjustment adoption. The results from these studies are summarised in Table 7.2, which indicates that female gender has a roughly equal number of significant and nonsignificant results. Education and income have more nonsignificant than significant results. Age and ethnicity have more nonsignificant results than positive results and even some negative correlations. Marital status, children in the home and—especially—homeownership and community tenure have too few results to even classify reliably.

Table 7.1 Summary of correlations with hazard adjustment, by independent variable and hazard

Hazard	Risk perceptions	Adjustment perceptions	Stakeholder perceptions	Hazard intrusiveness	Hazard experience	Hazard proximity
Earthquake	2 P 4 ns	3 P	3 P 1 ns	1 P	5 P 2 MP 1 ns	1 P 2 ns
Flood	3 P 1 ns	1 MP 1 ns	1 ns	–	2 P 1 MP	2 P
Hurricane	5 P	1 P	1 MP	–	1 P 4 MP 1 ns	1 P 1 MP
Tornado	2 P 2 MP	1 P 1 ns	2 P	2 P	1 P 2 MP	–
Volcano	1 P 2 MP 2 ns		1 P	1 P	1 P 1 MP	–
Total	13 P 2 MP 5 ns	5 P 1 MP 2 ns	6 P 1 MP 2 ns	4 P	10 P 9 MP 2 ns	4 P 1 MP 2 ns

P consistent positive effects, *MP* mixed positive and nonsignificant effects, *ns* nonsignificant effects, *MN* mixed negative and nonsignificant effects, *N* consistent negative effects

Table 7.2 Summary of correlations with hazard adjustment, by demographic variable and hazard

Hazard	Female gender	Age	Education	Income	White ethnicity	Married	Child	Owner	Tenure
Earthquake	2 P 2 ns	1 P 1 ns 1 N	1 P 1 ns	1 P 1 ns	2 ns	2 P 1 ns	1 P 1 ns	–	–
Flood	1 ns	1 P	1 ns	1 P 2 ns	–	–	–	–	–
Hurricane	2 P 2 MP 1 ns	2 P 1 MP 3 ns	1 MP 4 ns	2 P 3 ns 1 MN	1 P 1 MP 1 ns 1 MN	2 ns	2 MP	2 MP	1 P 1 MP
Tornado	1 ns	1 ns	1 P	1 ns	1 ns	–	–	–	–
Volcano[a]	–	–	–	–	–	–	–	–	–
Total	4 P 2 MP 5 ns	4 P 1 MP 5 ns 1 N	2 P 1 MP 6 ns	4 P 7 ns 1 MN	1 P 1 MP 4 ns 1 MN	2 P 3 ns	1 P 2 MP 1 ns	2 MP	1 P 1 MP

P consistent positive effects, *MP* mixed positive and nonsignificant effects, *ns* nonsignificant effects, *MN* mixed negative and nonsignificant effects, *N* consistent negative effects

[a]No demographic variables reported for volcano hazards

7.5 Conclusions

This chapter updates Lindell and Perry's (2000) review of research on the predictors of household hazard adjustment adoption by including the results of ten more recent North American studies on earthquakes and adding 16 studies on flood, hurricane, tornado, and volcano hazards. The results of these studies provide the basis for the theoretical conclusions and methodological recommendations that follow.

7.5.1 Theoretical Conclusions

The findings from this review are generally consistent with those from Lindell and Perry's (2000) review. First, these studies indicate that risk perceptions play a significant role in hazard adjustment. The majority of the results indicate a positive correlation with hazard adjustment adoption but there are some cases with mixed positive and nonsignificant results or even consistent nonsignificant results. One objective for future research is to explain why risk perceptions are not always significantly correlated with hazard adjustment adoption. One possibility is that the inconsistent results are due to variations in the operationalisation of this variable. As Lindell (2012) noted, risk has been measured in at least five distinctly different ways ranging from measures of hazard agent characteristics to affective reactions. However, differences in operationalisations cannot be the entire explanation because the research record shows that some notably different measures of risk perceptions have significant correlations with hazard adjustment adoption. Weinstein and Nicolich (1993) have provided another explanation—as people with high levels of risk perceptions adopt more hazard adjustments, their risk perceptions decrease. To investigate this possibility, future studies should either be longitudinal or, if they are cross-sectional, should include self-reports of current behaviour and intentions to adopt hazard adjustments in the future. Both longitudinal designs and cross-sectional designs that collect current behaviour and behavioural intention allow researchers to assess the relationship of risk perceptions with *changes* in behaviour.

Second, it is clear that hazard experience is a consistent predictor of hazard adjustment adoption. However, as is the case with risk perceptions, there are some mixed significant and nonsignificant results to be explained. Here too, there are differences in the operationalisation of this variable that might explain the variation in the results. In addition, there is the more interesting question of *how* hazard experience affects hazard adjustment adoption. One logical presumption is that the effect of hazard experience on hazard adjustment adoption is mediated by risk perceptions and there is some evidence that this is so (Lindell and Hwang 2008). However, Lindell and Prater (2000) found a correlation between hazard experience and hazard adjustment adoption that could not be mediated by risk perceptions because risk perceptions themselves had nonsignificant correlations with hazard adjustment adoption. Thus, another possibility is that the effect of hazard experience on adjustment adoption is mediated by adjustment perceptions or stakeholder perceptions or both.

Third, adjustment perceptions also appear to play a significant role in hazard adjustment adoption. Correlations involving this category of variables show more significant than nonsignificant results. However, support for this conclusion is weak because relatively few studies have included these variables. Thus, more research is needed to test the proposition of the PADM that risk perceptions motivate people to take protective action but adjustment perceptions determine whether any of the available alternatives are considered to be effective and feasible (Lindell and Perry 2004).

Fourth, stakeholder perceptions also have more positive than nonsignificant results but, like adjustment perceptions, support for this finding is relatively weak because so few studies have included these variables. As is the case with hazard experience, there are sound theoretical reasons (e.g. Theory of Reasoned Action, Theory of Planned Behaviour, Elaboration Likelihood Model) to believe that stakeholder perceptions could have either direct or indirect effects on the adoption of hazard adjustments. Future research needs to extend the findings of Arlikatti et al. (2007) by identifying the personal characteristics or situational conditions under which stakeholder perceptions have either direct or indirect effects.

Fifth, hazard intrusiveness appears to be a promising explanatory variable because all cases in which it was used yielded positive results. Once again, the conclusion must be regarded as tentative because it was tested so few times. Hazard intrusiveness—the frequency of thought and discussion about a hazard and adjustments to it—is likely to play an important role in the hazard adjustment process because long-term threats lack the specific deadline that is associated with imminent threats. As a consequence, procrastination can allow more immediate concerns of daily living (ranging from investments in home repairs to impulse purchases of tickets to sports events) to receive a greater priority than adjustments to environmental hazards unless frequent reminders cause those hazards to remain high on the household agenda. It might be possible to get more people to protect themselves and their loved ones from disaster by frequently reminding people that they are exposed to a hazard rather than trying to occasionally frighten them with messages about the severity of the consequences of exposure.

Sixth, hazard proximity has generally had a significant effect on hazard adjustment adoption but was only tested in a few studies. Tests of hazard proximity might have been infrequent because it was difficult to assess risk area residents' hazard proximity in the past. However, this impediment has been reduced by the increasing availability of Geographical Information Systems. Utilization of this resource should facilitate further tests of the hypothesized causal chain— "hazard proximity—hazard experience—risk perceptions—hazard adjustment" (cf. Lindell and Hwang 2008).

Finally, the inconsistent correlations of demographic variables with hazard adjustment in these studies replicates and extends Lindell and Perry's (2000) conclusion about demographic variables' lack of predictive validity for seismic hazard adjustment and is consistent with Baker's (1991) conclusion about demographic variables' lack of consistent predictive ability in hurricane evacuation. As is the case with hazard experience, the significant results that do exist for demographic variables might be attributable to having causal effects on variables such as hazard proximity, as proposed in the social vulnerability literature (Wisner et al. 2004). Specifically, this literature has concluded that the poorest segments of society are often forced to live in the most dangerous areas and in the most dilapidated housing because they cannot afford to live more safely.

7.5.2 Methodological Recommendations

This review has identified a number of methodological limitations in the research on household hazard adjustment. First, although some researchers have adopted—usually with modification—previous researchers' measures, there remains substantial diversity in the measures used for all variables. This is a practice that has both advantages and disadvantages. On the one hand, finding consistently significant correlations despite multiple operationalisations provides evidence of robustness of the results. On the other hand, multiple operationalisations make it difficult to make comparisons across studies. This is compounded by the problem that few studies report the average level of hazard adjustment. Both practices—changing measures and failures to report average levels of hazard adjustments—limit researchers' ability to identify differences among communities, differences among hazards, and changes over time in the levels of hazard adjustment adoption.

Another significant problem is the failure to report a matrix of intercorrelations among all the variables that were measured. The unavailability of such correlations impedes the replacement of narrative literature reviews, such as this one, with meta-analyses (Borenstein et al. 2009) that can assess the average magnitude of the effects of independent variables on hazard adjustment and the intercorrelations of the independent variables with each other. In some cases, researchers have reported only the test statistics and their significance levels (e.g. χ^2 or t-statistics and their associated degrees of freedom and p values). Such test statistics are a function of the strength of association *and the number of degrees of freedom*, so these test statistics can be extremely large (and have very small p values) if the sample size is very large even if the magnitude of the effect is extremely small (e.g. accounting for less than 1% of the variance in the dependent variable). Failure to report correlation matrices can preclude other researchers' assessments of the plausibility of regression models other than the ones that the original researchers tested. This is an important problem because many studies have only reported the results of stepwise regressions that identify the most parsimonious set of predictor variables. Stepwise regression analyses can produce extremely misleading results when, for example, two independent variables are correlated with each other but one has a slightly higher correlation with the dependent variable. Stepwise regression analysis will generally only allow the variable with the slightly higher correlation to enter the equation—even if that correlation is higher only because of sampling fluctuations. Thus, the stepwise regression analysis will suggest that one of the independent variables is an important predictor and the other is not when, in fact, both variables are equally important predictors.

Finally, research analyses that test single stage-single equation models cannot assess the validity of mediation models, such as the "hazard proximity—hazard experience—risk perceptions—hazard adjustment" model discussed above. Multistage—multiequation models are needed, as are longitudinal designs. If hazards researchers adopt these changes in reporting practices, research analyses, and research designs, future reviews will be able to extract much more valuable information from the studies that are published.

Acknowledgments This work was supported by the National Science Foundation under Grants SES 0527699, SES 0838654, CMM-0826401, and Grant CMM-1138612. None of the conclusions expressed here necessarily reflect views other than those of the author.

References

Arlikatti S, Lindell MK, Prater CS (2007) Perceived stakeholder role relationships and adoption of seismic hazard adjustments. Int J Mass Emerg Disaster 25(3):218–256

Baker EJ (1991) Hurricane evacuation behaviour. Int J Mass Emerg Disaster 9(2):287–310

Basolo V, Steinberg LJ, Burby RJ, Levine J, Cruz AM, Huang C (2009) The effects of confidence in government and information on perceived and actual preparedness for disasters. Environ Behav 41(3):338–364

Blanchard-Boehm RD (1998) Understanding public response to increased risk from natural hazards: application of the hazards risk communication framework. Int J Mass Emerg Disaster 16(3):247–278

Blanchard-Boehm RD, Cook MJ (2004) Risk communication and public education in Edmonton, Alberta, Canada on the 10th anniversary of the "Black Friday" tornado. Int Res Geogr Environ Educ 13(1):38–54

Blanchard-Boehm RD, Berry KA, Showalter PS (2001) Should flood insurance be mandatory? Insights in the wake of the 1997 New Year's Day flood in Reno–Sparks, Nevada. Appl Geogr 21(3):199–221

Borenstein M, Hedges LV, Higgins JT, Rothstein HR (2009) Introduction to meta-analysis. Wiley, Hoboken

Burton I, Kates R, White GF (1993) The environment as hazard, 2nd edn. Guildford Press, New York

CDRSS—Committee on Disaster Research in the Social Sciences (2006) Facing hazards and disasters: understanding human dimensions. National Academy of Sciences, Washington, DC

Chaiken S (1987) The heuristic model of persuasion. In: Zanna M, Olson J, Herman C (eds) The Ontario symposium. Erlbaum, Hillsdale, pp 3–39

Cohen J, Cohen P, West SG, Aiken LS (2003) Applied multiple regression/correlation analysis for the behavioral sciences, 3rd edn. Erlbaum, Mahwah

Drabek TE (1986) Human system responses to disaster: an inventory of sociological findings. Springer, New York

Eagly AH, Chaiken S (1993) The psychology of attitudes. Harcourt Brace, Fort Worth

Farley JE, Barlow HD, Finkelstein M, Riley L (1993) Earthquake hysteria before and after: a survey and follow-up on public response to the Browning forecast. Int J Mass Emerg Disaster 11(3):305–322

Faupel CE, Kelley SP, Petee T (1992) The impact of disaster education on household preparedness for Hurricane Hugo. Int J Mass Emerg Disaster 10(1):5–24

Federal Emergency Management Agency (2011) U.S. demographic shifts. www.fema.gov/pdf/about/programs/oppa/demography_%20paper_051011.pdf-2011-05-27. Accessed 20 Apr 2012
Fishbein M, Ajzen I (1975) Belief, attitude intention and behavior. Addison-Wesley, Reading
French JRP, Raven BH (1959) The bases of social power. In: Cartwright D (ed) Studies in social power. Institute for Social Research, Ann Arbor, pp 150–167
Garcia EM (1989) Earthquake preparedness in California. Urb Resour 5(4):15–19
Gladwin CH, Gladwin H, Peacock WG (2001) Modeling hurricane evacuation decisions with ethnographic methods. Int J Mass Emerg Disaster 19(2):117–143
Godschalk D, Parham D, Porter D, Potapchuk W, Schukraft S (1994) Pulling together: a planning and development consensus building manual. Urban Land Institute, Washington, DC
Heller K, Alexander D, Gatz M, Knight BG, Rose T (2005) Social and personal factors as predictors of earthquake preparation: the role of support provision, network discussion, negative affect, age, and education. J Appl Soc Psychol 35(2):399–422
Jackson EL (1977) Public response to earthquake hazard. Calif Geol 30(12):278–280
Jackson EL (1981) Response to earthquake hazard. Environ Behav 13(4):387–416
Kim Y-C, Kang J (2010) Communication, neighbourhood belonging and household hurricane preparedness. Disasters 34(2):470–488
Kramer RM (1999) Trust and distrust in organizations: emerging perspectives, enduring questions. Annu Rev Psychol 50:569–598
Lang JT, Hallman WK (2005) Who does the public trust? The case of genetically modified food in the United States. Risk Anal 25(5):1241–1252
Laska S (1990) Homeowner adaptation to flooding: an application of the general hazards coping theory. Environ Behav 22(3):320–357
Lindell MK (1994) Perceived characteristics of environmental hazards. Int J Mass Emerg Disaster 12(3):303–326
Lindell MK (2012) Response to environmental disasters. In: Clayton S (ed) Handbook of environmental and conservation psychology. Oxford University Press, New York, pp 391–413
Lindell MK, Hwang SN (2008) Households' perceived personal risk and responses in a multihazard environment. Risk Anal 28(2):539–556
Lindell MK, Perry RW (1992) Behavioural foundations of community emergency planning. Hemisphere Press, Washington, DC
Lindell MK, Perry RW (2000) Household adjustment to earthquake hazard: a review of research. Environ Behav 32(4):590–630
Lindell MK, Perry RW (2004) Communicating environmental risk in multiethnic communities. Sage, Thousand Oaks
Lindell MK, Perry RW (2012) The protective action decision model: theoretical modifications and additional evidence. Risk Anal 32(4):616–632
Lindell MK, Prater CS (2000) Household adoption of seismic hazard adjustments: a comparison of residents in two states. Int J Mass Emerg Disaster 18(2):317–338
Lindell MK, Prater CS (2002) Risk area residents' perceptions and adoption of seismic hazard adjustments. J Appl Soc Psychol 32(11):2377–2392
Lindell MK, Prater CS (2003) Assessing community impacts of natural disasters. Nat Hazard Rev 4(4):176–185
Lindell MK, Whitney DJ (2000) Correlates of household seismic hazard adjustment adoption. Risk Anal 20(1):13–25
Lindell MK, Alesch D, Bolton PA, Greene MR, Larson LA, Lopes R, May PJ, Mulilis J-P, Nathe S, Nigg JM, Palm R, Pate P, Perry RW, Pine J, Tubbesing SK, Whitney DJ (1997) Adoption and implementation of hazard adjustments. Int J Mass Emerg Disaster Spec Issue 15(3):327–453
Lindell MK, Prater CS, Perry RW (2006) Fundamentals of emergency management. Federal Emergency Management Agency Emergency Management Institute, Emmitsburg. www.training.fema.gov/EMIWeb/edu/fem.asp. Accessed 30 Mar 2012
Lindell MK, Arlikatti S, Prater CS (2009) Why people do what they do to protect against earthquake risk: perceptions of hazard adjustment attributes. Risk Anal 29(8):1072–1088

Mileti DS, Darlington J (1997) The role of searching in shaping reactions to earthquake risk information. Soc Probl 44(1):89–103

Mileti DS, Peek L (2000) The social psychology of public response to warnings of a nuclear power plant accident. J Hazard Mater 75(2–3):181–194

Mulilis JP, Duval TS (1997) The PrE model of coping with threat and tornado preparedness behavior: the moderating effects of felt responsibility. J Appl Soc Psychol 27(19):1750–1766

Mulilis JP, Lippa R (1990) Behavioral change in earthquake preparedness due to negative threat appeals: a test of protection motivation theory. J Appl Soc Psychol 20(8):619–638

Mullis JP, Duval TS, Rogers R (2003) The effect of a swarm of local tornados on tornado preparedness: a quasi-comparable cohort investigation. J Appl Soc Psychol 33(8):1716–1725

Nguyen LH, Shen H, Ershoff D, Afifi AA, Bourque LB (2006) Exploring the causal relationship between exposure to the 1994 Northridge Earthquake and pre- and post-earthquake preparedness activities. Earthq Spectra 22(3):569–587

NOAA—National Oceanic and Atmospheric Administration (2012) Weather ready nation: a vital conversation on tornadoes and severe weather. www.joss.ucar.edu/events/2011/weather_ready/outcomes.html. Accessed 5 May 2012

Noji EK (1997) The nature of disaster: general characteristics and public health effects. In: Noji EK (ed) The public health consequences of disasters. Oxford University Press, New York, pp 3–20

Norris FH, Smith T, Kaniasty K (1999) Revisiting the experience-behavior hypothesis: the effects of Hurricane Hugo on hazard preparedness and other self-protective acts. Basic Appl Soc Psychol 21(1):37–47

Nunnally JC, Bernstein IH (1994) Psychometric theory, 3rd edn. McGraw-Hill, New York

Palm R, Hodgson M, Blanchard R, Lyons D (1990) Earthquake insurance in California. Westview Press, Boulder

Peacock WG (2003) Hurricane mitigation status and factors influencing mitigation status among Florida's single-family homeowners. Nat Hazard Rev 4(3):149–158

Perry RW (1990) Volcanic hazard perceptions at Mt. Shasta. Environ Prof 12(4):312–318

Perry RW, Lindell MK (1986) Source credibility in volcanic hazard information. Volcano News 22(12):7–10

Perry RW, Lindell MK (1990) Predicting long term adjustment to volcano hazard. Int J Mass Emerg Disaster 8(2):117–136

Perry RW, Lindell MK (2008) Volcanic risk perception and adjustment in a multi-hazard environment. J Volcanol Geotherm Res 172(3–4):170–178

Petty R, Cacioppo J (1986) The elaboration likelihood model of persuasion. In: Berkowitz L (ed) Advances in experimental social psychology. Academic, New York, pp 123–205

Pijawka KD, Mushkatel AH (1991) Public opposition to the siting of the high-level nuclear waste repository: the importance of trust. Policy Stud Rev 10(4):180–194

Preston V, Taylor SM, Hedge DC (1983) Adjustment to natural and technological hazards: a study of an urban residential community. Environ Behav 15(2):143–164

Raven B (1965) Social influence and power. In: Steiner I, Fishbein M (eds) Current studies in social psychology. Holt, Rinehart and Winston, New York, pp 371–382

Raven B (1993) The bases of power: origins and recent developments. J Soc Issues 49(4):227–251

Russell L, Goltz JD, Bourque LB (1995) Preparedness and hazard mitigation actions before and after two earthquakes. Environ Behav 27(6):744–770

Sattler DN, Kaiser CF, Hittner JB (2000) Disaster preparedness: relationships among prior experience, personal characteristics, and distress. J Appl Soc Psychol 30(7):1396–1420

Siegel JM, Shoaf KI, Afifi AA, Bourque LB (2003) Surviving two disasters: does reaction to the first predict response to the second? Environ Behav 35:637–654

Slovic P, Fischhoff B, Lichtenstein S (2001) Facts and fears: understanding perceived risk. In: Slovic P (ed) The perception of risk. Earthscan, London, pp 137–153

Sorensen JH, White G (1980) Natural hazards: a cross cultural perspective. In: Altman I, Rapoport A, Wohlwill JF (eds) Human behavior and the environment: advances in theory and research. Plenum Press, New York, pp 89–95

Terpstra T, Lindell MK (in press) Citizens' perceptions of flood hazard adjustments: an application of the protective action decision model. Environ Behav

Turner R, Nigg J, Heller-Paz D (1986) Waiting for disaster. University of California Press, Los Angeles

Weinstein ND, Nicolich M (1993) Correct and incorrect interpretations of correlations between risk perceptions and risk behaviors. Health Psychol 12(3):235–245

Weinstein N, Lyon JE, Rothman AJ, Cuite CL (2000) Preoccupation and affect as predictors of protective action following natural disaster. Br J Health Psychol 5(4):351–363

Weyman AK, Pidgeon NF, Walls J, Horlick-Jones T (2006) Exploring comparative ratings and constituent facets of public trust in risk regulatory bodies and related stakeholder groups. J Risk Res 9(6):605–622

Whitney DJ, Lindell MK, Nguyen DH (2004) Earthquake beliefs and adoption of seismic hazard adjustments. Risk Anal 24(1):87–102

Wisner B, Blaikie P, Cannon T, Davis I (2004) At risk: natural hazards, people's vulnerability and disasters, 2nd edn. Routledge, London

Chapter 8
Community Understanding of, and Preparedness for, Earthquake and Tsunami Risk in Wellington, New Zealand

David Johnston, Julia Becker, John McClure, Douglas Paton, Sara McBride, Kim Wright, Graham Leonard, and Miriam Hughes

Abstract The city of Wellington, New Zealand's capital, is exposed to a wide range of potentially devastating impacts from various natural hazards. It is situated in one of the most active seismic regions in New Zealand, creating a significant earthquake risk. Another hazard to which it is exposed is that of tsunami from local and distant sources. Given the variety of hazards that Wellington faces, consideration of how the risks from such hazards can be reduced is necessary. Preparedness activities can be undertaken to try and reduce risk, with individual household preparedness forming one such activity. Motivating citizens to prepare can be a difficult task. Educators have often long assumed that if individuals are told about the risk of hazards, then they will begin to prepare; however, this is not usually the case. This is reflected in Wellington where traditional public education (i.e. information dissemination) has predominantly been undertaken to inform people about the risk of earthquakes and tsunami. Results from evaluation surveys show that in the wake of the public education campaigns, awareness of the risk is high, but levels of comprehensive preparedness low. Consequently it is apparent that risk perception does not usually link directly with preparedness, but is amplified or attenuated through a variety of individual,

D. Johnston (✉) • J. Becker • S. McBride • K. Wright • G. Leonard • M. Hughes
Joint Centre for Disaster Research, Massey University/GNS Science,
1 Fairway Drive, Avalon, Wellington 5010, New Zealand
e-mail: david.johnston@gns.cri.nz; j.becker@gns.cri.nz;
s.mcbride@gns.cri.nz; k.wright@gns.cri.nz; g.leonard@gns.cri.nz

J. McClure
School of Psychology, Victoria University, Kelburn Parade,
Wellington 6012, New Zealand
e-mail: john.mcclure@vuw.ac.nz

D. Paton
School of Psychology, University of Tasmania, Private Bag 50,
7001 Hobart, Launceston, TAS, Australia
e-mail: Douglas.Paton@utas.edu.au

H. Joffe et al. (eds.), *Cities at Risk: Living with Perils in the 21st Century*, Advances in Natural and Technological Hazards Research 33,
DOI 10.1007/978-94-007-6184-1_8, © Springer Science+Business Media Dordrecht 2013

social psychological and community factors. Such factors are important to the preparedness process and must be considered when developing public education programmes. Programmes should include a traditional information dissemination element to build awareness of the risk, as well as a more interactive community-based component to foster important factors such as critical awareness, self-efficacy, outcome expectancy, action coping, participation, engagement and trust.

8.1 Introduction

The city of Wellington is situated at the southern end of the North Island of New Zealand and is the nation's capital city. It is exposed to a wide range of potentially devastating impacts from various natural hazards and is situated in one of the most active seismic regions in New Zealand (Cousins et al. 2008). Although the city has been subject to significant earthquakes in the past (e.g., 1848, 1855 and 1942), it has experienced a relatively calm period over the past 70 years. As well as earthquakes, Wellington is also at risk from tsunami from both local and distant sources (Berryman 2005; Cousins et al. 2008). In historic times large South American earthquakes (in 1886, 1877, and 1960) have generated tsunami that have reached Wellington, but with only minor impacts. The most significant locally generated tsunami was from the 1855 Wairapapa earthquake (Mw 8.2), that produced a tsunami in Lyall Bay with heights of around 9–10 m. Of increasing concern are potential tsunami from the East Coast subduction zone. For the Wellington region, most of the east coast of the North Island and much of the South Island coast, this subduction zone event is the most likely source of a damaging tsunami (Berryman 2005; Wallace et al. 2009).

While Wellington has experienced a period of relative quiescence with respect to large-scale natural disasters, other regions within New Zealand have not been so fortunate. During 2010–2011 the Canterbury region in New Zealand was subject to a series of destructive earthquakes that caused damage, disruption, injury, death and economic loss. Two main earthquakes of note as part of the sequence include the magnitude 7.1 Darfield earthquake which occurred on 4 September 2010, and the magnitude 6.3 Christchurch earthquake that occurred on 22 February 2011. The Darfield earthquake caused widespread damage and disruption to structures, services and livelihoods. The Christchurch earthquake also caused widespread damage and disruption, but was more serious in that it brought about the death of 185 people in Christchurch city, mostly related to building collapse and falling objects (New Zealand Police 2012). The series of earthquakes experienced by people in Canterbury and Christchurch is a timely reminder that disasters such as earthquakes and tsunami can happen at any time, and that many cities remain at risk around the world. Wellington is one of those cities.

Although earthquakes and tsunami are uncontrollable events, their impacts and consequences can be influenced by preparedness activities. Some preparedness takes the form of building regulations and other legislation. However, regulation also needs to be complemented by citizens' own preparatory efforts, such as retrofitting their houses, storing essential items necessary for survival after a disaster, or making emergency plans (Paton and Johnston 2006). Yet despite the risk of huge losses from earthquakes, many citizens and organisations do not make sufficient effort to prepare (Ministry of Civil Defence and Emergency Management 2009). The reasons for not preparing are many and are influenced by a variety of individual, social and environmental factors. It is therefore useful to clarify psychological and social factors that contribute to a failure to prepare for earthquakes, and to show how best we can overcome these obstacles.

The aim of this chapter is to outline the dynamics of risk perception, discuss how risk perception interacts with other psychological and social factors and outline how this overall process links to preparedness. The implications of this process for public education are discussed using Wellington city as an example.

8.2 Psychological and Social Factors Contributing to a Failure to Prepare

8.2.1 Risk Perception

While risk perception has been identified as a factor that contributes to people's decision-making about whether or not to prepare for disasters, its influence is often not straightforward or predictable from expert analyses. This reflects the fact that estimates of risk by experts are mostly based on objective analyses of the likelihood of hazard activity and its consequences within a specific area, whereas the estimates made by citizens, particularly when dealing with risk associated with uncertain natural hazards, are more subjective and socially constructed. For example, research in Napier (New Zealand) has identified how interactions with other community members play a key role in how people perceive risk (Mclvor et al. 2009). Disparities are thus common between these expert assessments and the manner in which they are interpreted and acted on by the public and other groups, including some local authorities (Paton et al. 2005; Paton 2006). This difference frequently remains even when people are directly presented with scientific information. This discrepancy is due to the fact that people's understandings of and responses to risk are determined not only by scientific information or direct physical experiences, but also by psychological, social, cultural, institutional and political processes that interact to impose meaning on uncertainty (Burns et al. 1993; Paton 2008; Spittal et al. 2008).

The different factors that affect risk perception are usually not independent, and can vary between different hazards. For example, an earthquake might be seen as a greater risk than a flood by an individual because it can happen without any warning, in contrast with flooding for which a warning can be given. This in turn may affect people's preparedness, with people in some cases more likely to prepare for an earthquake because of uncertainty surrounding the timing of the event, and less likely to prepare for a flood because they believe they can make preparations when they get a warning (Becker et al. 2012a).

Beliefs about risk and risk reduction behavior are also influenced by several social and cognitive biases. Two processes specifically relevant here are *unrealistic optimism* and the *normalisation bias* (Paton et al. 2000). Unrealistic optimism, sometimes referred to as the "illusion of invulnerability" is seen where people underestimate the risk to themselves and overestimate the risk to others (Weinstein and Klein 1996). Thus, while people may acknowledge objective risk in their community, they are more likely to attribute its negative implications to others rather than themselves. This bias leads people to take risky options. In some countries (e.g. USA and New Zealand where preparing is commonly seen as an individual rather than collective responsibility) citizens often think they are better prepared than others, which leads them to think that they will be safe (Helweg-Larsen 1999; Spittal et al. 2005).

The normalisation bias results when people extrapolate a capability to deal with major earthquakes from their experience of successfully enduring minor, rarely occurring earthquakes. Like the optimistic bias, this process results in people underestimating risk (relative to scientific and planning estimates) and acting in ways that, from an objective rationalistic perspective, are seen as counterintuitive and counterproductive (Johnston et al. 1999; Mileti and O'Brien 1992; Russell et al. 1995).

Denial of risk is a related bias that inhibits positive actions. Importantly, denial is often greatest among people who are most at risk, partly because denial serves to reduce anxiety. Lehman and Taylor (1987) showed that students living in dormitory buildings with poor seismic resistance in Los Angeles denied the seriousness of earthquake risk more than those who lived in sound buildings. Denial can be reduced by increases in people's perceptions of control over a hazard (Lehman and Taylor 1987).

In addition, people may tend to overestimate the capacity of hazard mitigation strategies to eliminate a threat through the operation of an interpretive bias known as *risk compensation* (Adams 1995). This phenomenon has also been known as the "levee syndrome". This construct describes how people maintain a balance between the perceived level of safety proffered by their environment and the level of risk manifest in their actions and attitudes. Thus, a perceived increase in extrinsic safety (e.g., hazard monitoring, structural mitigation) can decrease an individual's or group's perceived risk, thereby reducing the perceived need for action. This becomes problematic because emergency managers, in the process of disseminating information about mitigation options, assume that people's risk estimates, and thus their behaviours, remain constant and are not affected by the new information. This assumption is unfounded. The dissemination of information on mitigation and preparedness options has been found to lead to a reduction in levels of household and individual preparedness, and a transfer of responsibility for safety to civic authorities (Paton et al. 2000).

Misperceptions of risks not only come about because of the cognitive tendencies of an individual's mind, but can also be brought about during the communication process. First, information presented by scientists and officials may not always be clear enough for the public to understand, or at worst inaccurate. Additionally, the ways in which risk information is communicated can have a distorting effect on risk perceptions. For example, while risks with relatively minor bio-physical consequences can elicit extreme concern (such as burglary), very significant risks (such as the effects of natural disasters) can be underestimated by communities and organisations. This phenomenon is encompassed under "the social amplification of risk" framework (Kasperson et al. 1988). Public accusations of "irresponsible media", "organisational incompetence", and "public hysteria" are common when hazard outcomes differ from citizens' expectancies (Rip 1988). The objection generally arises when sources such as the media are seen to overemphasise adverse or catastrophic aspects of a problem and fail to provide a balanced view. Situations are further complicated where there is a lack of trust in information sources, particularly when these sources dismiss the concerns, needs, and interests of the community (Johnston et al. 2005).

8.2.2 Outcome Expectancy and Fatalism

People differ in their outcome expectancies about the value of preparedness actions (i.e. their beliefs about the effectiveness of preparing) (McClure et al. 1999). A key factor affecting outcome expectancy is fatalism, the attitude that "nothing that I do will make any difference, so there is no point in trying". In regard to earthquakes, fatalism relates to the belief that earthquakes are so powerful that there is no use preparing for them. When the "big one" comes, it will be so strong that the best of efforts will be laid to waste. This fatalism often reflects a failure to distinguish between the uncontrollable force of earthquakes and the relative controllability of their consequences. Damage from earthquakes can, for example, be attributed to both the intrinsic power of the earthquake and the structure of the buildings that collapse and kill people (McClure et al. 2001). Fatalistic representations acknowledge only one of these causes – the earthquake's uncontrollable force. Less fatalistically-oriented people tend to attribute damage to the combined effects of the earthquake and building design.

Research has shown that there are ways of reducing earthquake fatalism. One strategy is to show people that the damage that occurs in earthquakes is often distinctive in terms of the buildings that collapse (McClure et al. 2001) – for example, one house in a street may collapse while all the other houses stand firm. When people see that damage in earthquakes is distinctive in this way, they are more likely to attribute the damage to building design. News reports of earthquakes usually take the opposite tack, however, focusing on widespread damage rather than buildings that stood firm. An American reporter arriving at Kobe 2 days after the earthquake in 1995 said he expected to see the city devastated but he was amazed that all around him were buildings in good shape (Cowan et al. 2002). News reports immediately after an earthquake focus on the greatest damage, with headlines like "Kobe devastated". By contrast, later reports such as anniversary reports tend to be more analytic and focus on the characteristics of the buildings that collapsed and on lessons that can be learned. One study found that people who read these more analytic reports about the Kobe earthquake were less fatalistic about the damage than people who read the immediate news reports about the same earthquake (Cowan et al. 2002). The information to which citizens are exposed can therefore alter their fatalism and understanding.

Taking this strategy one step further, spelling out the specific features of building design that reduce damage helps people to see that something can be done to mitigate the effects of earthquakes (McClure et al. 2007). Of course, some aspects of building design that affect resilience are expensive, which is a disincentive to action, but this is not the case for all measures. For example, the use of simple computer locks was a leading predictor of business survival after the 1989 San Francisco earthquake, yet this precaution costs only a few dollars. Focusing people on the effectiveness of specific preventive actions such as this can help to undermine fatalistic beliefs about earthquake outcomes (Turner et al. 1986).

8.2.3 Anxiety

Individuals may experience anxiety when thinking about potential disasters such as earthquakes and tsunami. Anxiety may have positive or negative implications depending on the level of anxiety (Paton et al. 2005, 2003). High levels of anxiety may act as a hindrance to people's desire to get prepared (Paton et al. 2005, 2003). It is possible that those people with high levels of anxiety tend to allay their anxiety by denying that there is a problem, becoming overly-optimistic that they won't be affected by it, or becoming fatalistic about the risk (McClure 2006). Individuals that have lower levels of anxiety are likely to have enough concern that they want to prepare, but not so much that it leads to denial, over-optimism or fatalism (Paton et al. 2005, 2003). A number of earthquake researchers have found that a degree of anxiety or concern can assist with motivating preparedness (Dooley et al. 1992; Heller et al. 2005; Karanci and Aksit 2000; Karanci et al. 2005; Kiecolt and Nigg 1982; Rüstemli and Karanci 1999; Showalter 1993; Siegel et al. 2003; Turner 1983).

8.2.4 Motivation

Changing risk perceptions alone will not necessarily bring about behaviour change or direct action to address a particular risk. Social cognitive processes mediate between perceived risk and risk reduction actions, and whether a risk perception will be translated into corresponding behavior depends on these processes (Paton et al. 2006). *Motivation* is a particularly important mediating factor. People may not be motivated to take preparatory action if they do not accept their at-risk status or do not perceive hazards as salient. Irrespective of the level of risk, action is constrained if people see hazard effects as insurmountable (low outcome expectancy), see themselves as lacking the competence to act (low self-efficacy), or are not disposed to action (low action coping). Risk perception may not guide actions if people lack resources for implementation (low response efficacy), transfer responsibility for their safety to others (low perceived responsibility), lack trust in information sources, or stress uncertainty regarding the likely timing of hazard occurrence (Paton 2006).

People will be more motivated to prepare if they perceive that preparedness will make a difference to their outcome in a disaster (e.g. they will be more likely to survive the impact of the disaster itself, or will be better placed to survive and prosper during disaster recovery). They will also be more motivated to prepare if they perceive that they have a individual responsibility to prepare, or they perceive that they have a responsibility for others (e.g. children, elderly parents, employees, animals) (Becker et al. 2012a; Mclvor et al. 2009; Mulilis and Duval 1995, 1997; Perry and Lindell 2008).

Motivation of individuals to prepare will differ depending on the person, with some individuals being easier to motivate than others. Some may only require the right sort of information to motivate them, while for others more assertive strategies will need to be employed that enhance motivation. For example, some people may have made a conscious decision not to prepare, and these people will require strategies that get them to a point where they are motivated to think about preparing (e.g. strategies that individualise the hazard issue or show what other people have done). Others may be interested in preparing already, and only require specific information (e.g. about their vulnerability and the severity of event they could experience) to enhance their motivation for taking preparedness action (Paton and McClure, in preparation).

8.2.5 Model of Resilience

One basic way of representing risk is that it is the product of estimates of likelihood and hazard consequences. Risk associated with the "consequences" side of this equation is the focus of much risk communication research and practice that seeks to facilitate preparedness and resilience. Research into people's preparedness, and thus their capacity to manage their risk by reducing their exposure to adverse consequences, has identified that decisions reflect people's interpretation of risk and information about risk. When dealing with uncertain events, decisions are based on the interaction between interpretative processes at individual-, community- and societal-level sources, with trust having a pivotal role in decision making under uncertainty (Paton 2008; Becker et al. 2012a).

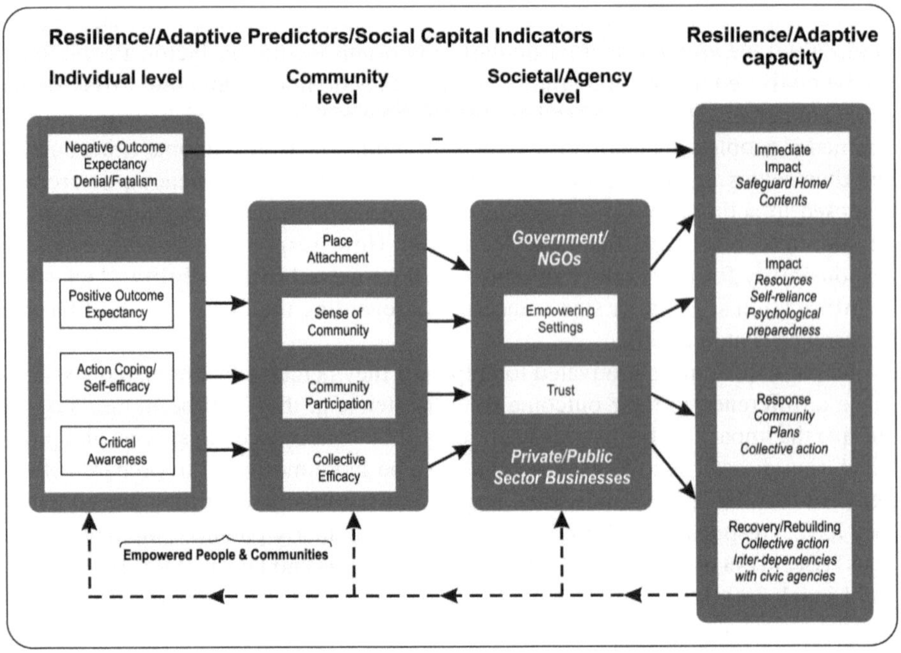

Fig. 8.1 A model of community resilience, showing selected indicators at each level (individual, community and societal/agency), different types of resilience (including household preparedness) and relationships between them (Paton 2010, 2012)

Recent research in Australasia has developed a generic model of community resilience (Fig. 8.1), which takes into account many of the aforementioned psychological and social factors that influence whether people prepare for disasters or not. This model depicts many of the factors contributing to resilience, and the interaction (pathways) between them (Daly et al. 2009; Paton 2006). The factors of most relevance can be grouped into three areas: (1) *individual*, where people are confident that their actions can make a positive difference for themselves, their families and their neighbours (outcome expectancy and action coping); (2) *community*, where residents actively participate in their communities to identify and discuss their issues and risks and determine collective solutions (community participation and problem articulation); and (3) *societal/agency*, where communities are supported by civic agencies that encourage and empower community-led initiatives and where mutual trust and respect exist (empowerment and trust).

The contents of Fig. 8.1 also identify the need for preparedness (resilience/adaptive capacity) to include not only the safeguarding of homes and emergency kits, but also developing psychological preparedness, community planning, and the development of functional relationships between communities and civic agencies (Paton 2006, 2012). The contents of Fig. 8.1 can inform the development of intervention through public education and community outreach.

8.3 The Role of Public Education in Influencing Risk Perception, Resilient Factors and Preparedness

Public education can be used to address many of the psychological and social factors that influence people's preparedness and overall resilience. Education programmes can include appropriate messages and participatory activities that develop helpful attitudes toward risk and build the factors that contribute to resilience. Most traditional public education programmes have tended toward having a focus on information dissemination via various forms of media, with a focus on raising awareness of the risk of natural hazards and of the need to prepare. Public education to date has had less of a focus on building other factors that are important to preparing such as coping style, participation, empowerment and trust (Becker et al. 2012b).

Public education about natural hazards in Wellington is delivered both at a national and local level. The New Zealand National Public Education Strategy sets out the strategic framework for public education of the Emergency Management sector in New Zealand for the 2006–2015 period. The national public education programme is a traditional form of public education which relies predominantly on information dissemination via media. The programme has multiple elements including: (1) media advertising (television, radio and print); (2) advertising in the "Yellow Pages" regional directories; (3) dedicated website (www.getthru.govt.nz); (4) printed brochures; (5) household mail drop (with emergency plan and check list); (6) promotional display stands and drink bottles; (7) a "Disaster Awareness Week"; (8) school resources ("What's The Plan Stan?" – see www.whatstheplanstan.govt.nz); (9) public relations, sponsorship and promotional activities; (10) online Civil Defence Emergency Management (CDEM) public education toolbox (Ministry of Civil Defence and Emergency Management 2006a). Within the Wellington Region, the CDEM Group plan outlines a range of public education initiatives delivered at a local and regional level. Current efforts are based around a mix of media advertising, printed material, public outreach and community-based programmes in schools and community groups (Wellington Region Emergency Management Group 2005).

Evaluating the effectiveness of public education presents a number of challenges. The national programme has been accompanied by an evaluation programme. The focus of the evaluation is to acquire information on national levels of awareness and preparedness (Ministry of Civil Defence and Emergency Management 2006b, 2007, 2008a, 2009). Between 2006 and 2012, seven annual surveys were undertaken to measure people's disaster preparedness and to assess the effectiveness of the national campaign. Using a 29-item measure, telephone interviews were conducted with 1,000 randomly selected people nationwide. The Wellington sample was between 112 and 118 people across the seven surveys.

The surveys asked people about their awareness and understanding of the types of disaster that could occur in Wellington. With respect to earthquakes, awareness of the earthquake risk in Wellington was very high, with 95% of people aware of the earthquake risk in 2006; 98% in 2008; and 98% in 2010. This is comparable to nationwide figures, with over 90% of people thinking that an earthquake could affect them in their lifetime. Nationwide tsunami awareness was also reasonably high with over 70% of individuals believing a tsunami could affect them.

In a 2003 survey of earthquake and tsunami awareness of coastal residents in Wellington, only between 40 and 50% of residents reported thinking it likely an earthquake would affect their community, and only 10–35% reported thinking it likely a tsunami would do the same (Johnston et al. 2003). Thus it appears that the traditional public education programmes that have been run since the 2003 survey have been influential in raising awareness and risk perception. The Canterbury earthquakes have had little impact on raising the level of earthquake risk perception in Wellington as awareness was already high before the earthquake sequence took place, and surveys in 2011 show that this perception remains at a high level post-earthquake (McClure et al. 2011). This is not to say that there has been no influence of the Canterbury earthquakes on the risk perception of Wellingtonians, however, as changes may have taken place that are related to other risk-related perceptions (e.g. people may already be aware that Wellington has an earthquake risk, but post- Canterbury earthquakes be more likely to perceive that an earthquake could happen to them, any time).

With respect to preparedness, the figures from the national evaluation programme tell a slightly different story. Table 8.1 presents the proportion of the Wellington sample from the programme that fell into the categories of (1) fully prepared[1]; (2) prepared at home[2]; (3) has a plan; (4) has survival items.

The statistics from the evaluation programme show that in general, less than a third of people in Wellington are fully prepared for a disaster. Slightly larger proportions (between a quarter and half) are prepared at home. Around half to three quarters of people have said that they have made a plan about what to do in a disaster, and over three quarters of people have some survival items at home. These figures are consistent with other studies that report low levels of full preparedness, and higher numbers of individuals who retain some survival items (Paton et al. 2005). Some aspects of preparedness in Wellington have likely been influenced by the Canterbury earthquakes of 2011 and 2012 (for example, being prepared at home rose from 34 to 51% between 2010 and 2011; and people were more likely to report that they had developed a household emergency plan post-2010), but overall people are still not fully prepared for a disaster with continuing low levels of full preparedness in 2011 (33%) and 2012 (25%) (Ministry of Civil Defence and Emergency Management 2012).

Despite some minor changes in preparedness over the last 7 years, there is little evidence of any significant increase. It appears that current public education programmes have been effective at raising awareness about the risks posed by hazards such as earthquakes and tsunami, but have done little to enhance preparedness. This finding is consistent with many other overseas studies and suggests that having an

[1] Fully prepared is defined in the survey as being prepared both at work and home by (1) having an emergency survival plan; (2) having emergency items and water; and (3) undertaking regularly updated emergency survival items.

[2] Prepared at home is defined in the survey as (1) having an emergency survival plan; (2) having emergency items and water; and (3) undertaking regularly updated emergency survival items.

Table 8.1 Preparedness levels in Wellington from the Campaign Monitoring Research (Ministry of Civil Defence & Emergency Management 2006a, b, 2007, 2008a, b, c, 2009, 2010, 2011, 2012)

	2006 (*N* not reported for Wellington), %	2007 (*N*=114), %	2008 (*N*=118), %	2009 (*N*=113), %	2010 (*N*=113), %	2011 (*N*=112), %	2012 (*N*=109), %
Fully prepared[1]	18	16	24	14	18	33	25
Prepared at home[2]	35	37	41	26	34	51	40
Has a plan	59	67	63	49	57	79	71
Has survival items	76	77	82	71	86	88	83

accurate risk perception about a hazard will not necessarily lead to action (Solberg et al. 2010). The link between risk perception and preparedness appears to be indirect and mediated by other psychological and social factors (Whitney et al. 2004).

It is important to be aware of the risks posed by a hazard, as people must understand the risks before they have a motivation to prepare (Paton et al. 2006). Traditional means of public education (such as the information dissemination strategies used in the national programme) can assist with this awareness and motivation. Such programmes help to create an awareness of the hazardscape, and understanding of the risks posed, initiate discussion and thinking about hazards and their repercussions, and instill an understanding of how certain actions will be of benefit (Paton et al. 2006). However, other strategies are also needed to help develop additional factors that are important for influencing preparedness (e.g. self-efficacy, outcome expectancy, action coping, sense of community, participation, empowerment and trust).

Possibilities exist for traditional public education initiatives to be augmented with other measures to increase their efficacy. Community development and community-based education programmes (e.g., based on models such as that described in Fig. 8.1 that identify key elements of community engagement strategies) have been found to effectively increase earthquake and tsunami preparedness and foster community empowerment (Finnis 2004, 2007; Finnis et al. 2007; Said et al. 2011). These programmes also promote qualities in people that increase resilience to disasters like earthquakes and tsunami, such as self-efficacy, using action coping mechanisms and a sense of community. Such qualities cannot be developed through information dissemination alone.

In Wellington a variety of such programmes exist, ranging from the conventional concept of a community-based public education project where education is issued at a community level, to projects that involve physical modifications, community action projects, and action research. Although not all of the programmes are specifically focused on providing "hazard education" (e.g. some are related to wider sustainability and environmental management issues), such programmes are valuable as they undertake beneficial physical modifications, bring about change, and/or empower communities and their residents by building on the qualities necessary to motivate preparedness. These qualities include sense of community, self-efficacy, problem-focused coping, response efficacy and outcome expectancy (Paton 2006).

A number of community-based emergency management groups operate in Wellington under the oversight of the Wellington Region Emergency Management Office (WREMO). One such community-based emergency management group is the Civil Defence in Karori, which is a well established group with numerous volunteers undertaking many activities and community education events. The group is supported by the local community constable, who is responsible for the Neighbourhood Support programme which he heavily promotes in conjunction with Civil Defence. Community Patrol, an active community policing group in Karori, also works closely with the group. Civil Defence information and activities are promoted mostly through school fairs, but also through presentations at local community groups and club meetings, such as the Bridge Club, Probus and sporting groups. Churches are also engaged and the Civil Defence Centre (CDC) manager and police constable has been invited to address congregations as part of the Sunday sermon. The CDC has also established memoranda of understanding with the local supermarkets, pharmacies, recreation centre, and other vital stores and services for the requisition of goods and services in an emergency. Further, the Karori CDC is now working closely with all schools in the area by providing them with radios that are tested daily by the schools, students and teachers by using them at cross patrols, making relay contact between the schools, and making contact with the police station. The success of Karori CDC activities is thought to be due to the nature of the suburb. It is relatively isolated, has a low turnover of residents and generations of families have lived there and return to live there. Residents already feel a sense of community and are prepared to participate. CDC is helping people to realise just how alone Karori, in particular, will be in the event of a large earthquake, and how important it is that they help each other to prepare. It is also developing attributes such as critical awareness, self-efficacy, positive outcome expectancy, action coping, community participation, and trust in officials and each other, and is empowering residents to prepare for a disaster.

Another example of a project with strong community involvement is the tsunami evacuation mapping project completed for the wider Wellington Region in 2009 (Leonard et al. 2009). This project aimed to define evacuation zones for all tsunami sources from the local subduction zone through the South Pacific to distant sources such as Peru and Chile. Pilot community-based evacuation maps and signage were developed for the suburb of Island Bay in 2010/11, with roll-outs underway for 12 more areas from 2012. The signs and maps follow national standards and guidelines respectively (MCDEM 2008b, c), but focus heavily on community response planning, community decisions over evacuation routes, signage types and placement, and community lead evacuation exercises. Blue lines were painted across all streets at the inland edge of the largest evacuation zone to denote the point beyond which people should evacuate in all tsunami, especially for the largest local-source subduction event. The blue lines were a community idea, and the pilot project was observed for over a year to evaluate the reaction of the wider community and city population. Generally positive responses and very widespread awareness of the lines and of the correct evacuation actions have led to a roll-out throughout the city, along with international interest in the concept. Such a community-based project has not only raised awareness about the tsunami hazard in Wellington and the need to prepare for taking the correct actions, but has also helped develop attributes in community members related to self-efficacy, positive outcome expectancy, articulating and solving problems, participation, empowerment and trust.

8.4 Conclusions

Recent hazard modeling has improved understanding of the risk to Wellington from earthquake and tsunami events. Events from the Wellington Fault (Cousins et al. 2008) or the east coast subduction zone have the potential to result in major loss of life and substantial damage to Wellington (Wallace et al. 2009). Therefore there is a compelling and urgent need to develop more effective strategies for reducing the risk posed by future disasters in Wellington and other parts of New Zealand.

Being prepared for a disaster is one way of reducing the risk, and individual household preparedness is one sub-set of preparedness that should be employed. It is often assumed that traditional education programmes that tell people about the risk will raise people's awareness and risk perception and directly motivate them to prepare. This assumption is unfounded however, which is reflected in data for Wellington. Traditional information dissemination programmes that have been run in the city have been successful in raising risk perception about earthquakes and tsunami in Wellington, but levels of full preparedness are still low amongst the population.

Research has revealed some answers about why there does not appear to be a direct link between risk perception and preparedness. Risk perception is actually amplified or attenuated through a variety of individual, social psychological and community factors, as part of people's decision-making process about whether to prepare or not. At an individual level, biases such as the unrealistic optimism bias, normalisation bias and risk compensation may come into play. Anxiety, fatalism and denial of risk may also affect people's perceptions about risky environments and their belief in the benefits of preparing. Those who perceive that preparing will make a difference to their outcome in a disaster will be more inclined to become prepared. Other factors that contribute to people's motivation and ability to prepare include self-efficacy, a propensity to undertake action coping, a belief that they have a responsibility to prepare, a feeling of responsibility for others, participation in society, a sense of community, a feeling of empowerment, and trust in information and officials.

Consequently public education should not solely be focused on raising people's risk perception of hazards; it should also account for the numerous other factors that interact with risk perception. Traditional information dissemination-based methods of public education can be used to raise awareness, and get individuals thinking and talking about hazards and preparedness. Information can also be used to relate specific messages that address biases, reduce denial and fatalism and engender positive outcome expectancy. However, traditional public education should also be coupled with more interactive community-based components. The active involvement of communities in hazard education and other mitigation activities has been shown to be effective in enhancing preparedness (Finnis 2007; Finnis et al. 2007; Paton 2005). Community-based programmes can develop the qualities in people (e.g. self-efficacy, action coping, sense of community) that promote preparedness and increase resilience to disasters. Wellington has begun this process in both in the context of earthquake and tsunami hazards, and provides a good example of how education programmes should be evolving.

Acknowledgements David Johnston and Julia Becker gratefully acknowledge EPICentre for supporting their attendance at the Cities at Risk workshop in London in September 2009. The research was also part funded by the New Zealand Foundation for Research Science and Technology under contract No. C05X0402.

References

Adams J (1995) Risk. UCL Press, London

Becker JS, Paton D, Johnston DM, Ronan KR (2012a) A model of household preparedness for earthquakes: how individuals make meaning of earthquake information and how this influences preparedness. Nat Hazards 64(1):107–137

Becker JS, McBride S, Paton D (2012b) Improving community resilience in the Hawke's Bay: a review of resilience research, and current public education, communication and resilience strategies. GNS Science report, Lower Hutt

Berryman KR (2005) Review of tsunami hazard and risk in New Zealand. Institute of Geological & Nuclear Sciences client report 2005/401. http://www.civildefence.govt.nz/memwebsite.nsf/Files/Tsunami_Hazard_report/$file/Final_Hazard_and_Risk_Report_part_1.pdf. Accessed 10 Oct 2012

Burns WJ, Slovic P, Kasperson RE, Kasperson JX, Renn O, Emani S (1993) Incorporating structural models into research on social amplification of risk: implications for theory construction and decision making. Risk Anal 13(6):611–623

Cousins J, Power W, Destegul U, King A, Trevethick R, Weir B, Blong R, Miliauskas B (2008) Earthquake and tsunami losses from major earthquakes affecting the Wellington region. http://www.aon.com/attachments/reinsurance/200812_ab_analytics_nz_tsunami.pdf. Accessed 28 Mar 2011

Cowan J, McClure J, Wilson M (2002) What a difference a year makes: how immediate and anniversary media reports influence judgments about earthquakes. Asian J Soc Psychol 5(3):169–185

Daly M, Becker J, Parkes B, Johnston D, Paton D (2009) Defining and measuring community resilience to natural disasters. Tephra 22:15–20

Dooley D, Catalano R, Mishra S, Serxner S (1992) Earthquake preparedness: predictors in a community survey. J Appl Soc Psychol 22(6):451–470

Finnis KK (2004) Creating a resilient New Zealand: can public education and community development campaigns create prepared communities? An examination of preparedness motivation strategies. Ministry of Civil Defence and Emergency Management, Wellington

Finnis KK (2007) Community-based public education initiatives. Joint Centre for Disaster Research Report 2007/01. http://tur-www1.massey.ac.nz/~psyweb/disasters/pubs/JCDR/JCDR%20Research%20Report%202007%2001%20Finnis_Community-based%20public%20education%20initiatives.pdf. Accessed 10 Oct 2012

Finnis KK, Johnston DM, Becker JS, Ronan KR, Paton D (2007) School and community-based hazards education and links to disaster-resilient communities. Reg Dev Dialogue 28(2):99–108

Heller K, Alexander DB, Gatz M, Knight BG, Rose T (2005) Social and individual factors as predictors of earthquake preparation: the role of support provision, network discussion, negative affect, age, and education. J Appl Soc Psychol 35(2):399–422

Helweg-Larsen M (1999) (The lack of) optimistic bias in response to the Northridge earthquake: the role of individual experience. Basic Appl Soc Psychol 21(2):119–129

Johnston D, Bebbington MS, Lai C-D, Houghton BF, Paton D (1999) Volcanic hazards perceptions: comparative shifts in knowledge and risk. Disaster Prev Manag 8(2):118–126

Johnston D, Leonard G, Bell R, Stewart C, Hickman M, Thomson J, Kerr J, Glassey P (2003) Tabulated results of the 2003 National Coastal Community Survey. Institute of Geological & Nuclear Sciences, Lower Hutt

Johnston D, Paton D, Crawford G, Ronan K, Houghton B, Bürgelt PT (2005) Measuring tsunami preparedness in coastal Washington, United States. In: Bernard E (ed) Developing tsunami-resilient communities. Springer, Dordrecht, pp 173–184

Karanci AN, Aksit B (2000) Building disaster-resistant communities: lessons learned from past earthquakes in Turkey and suggestions for the future. Int J Mass Emerg Disasters 18(3):403–416

Karanci AN, Aksit B, Dirik G (2005) Impact of a community disaster awareness training program in Turkey: does it influence hazard-related cognitions and preparedness behaviors? Soc Behav Individual 33(3):243–258

Kasperson RE, Renn O, Slovic P, Brown HS, Emel J, Goble R, Kasperson JX, Ratick S (1988) The social amplification of risk: a conceptual framework. Risk Anal 8(2):177–187

Kiecolt KJ, Nigg JM (1982) Mobility and perceptions of a hazardous environment. Environ Behav 14(2):131–154

Lehman DR, Taylor SE (1987) Date with an earthquake: coping with a probable, unpredictable disaster. Individual Soc Psychol Bull 13(4):546–555

Leonard GS, Power WL, Lukovic B, Smith WD, Langridge RM, Johnston DM, Downes GL (2009) Interim tsunami evacuation planning zone boundary mapping for the Wellington and Horizons regions defined by a GIS-calculated attenuation rule, GNS Science Report 2008/30. GNS Science, Lower Hutt

McClure J (2006) Guidelines for encouraging householders' preparation for earthquakes in New Zealand. Report for building research. http://www.adpc.net/v2007/programs/CBDRM/PROGRAMS/PDRSEA4/Echoes/guidelines%20for%20encouraging%20household%27s%20preparation%20for%20EQ%20in%20NZ.pdf. Accessed 10 Oct 2012

McClure J, Walkey F, Allen M (1999) When earthquake damage is seen as preventable: attributions, locus of control and attitudes to risk. Appl Psychol Int Rev 48(2):239–256

McClure JL, Allen M, Walkey FH (2001) Countering fatalism: causal information in news reports affects judgments about earthquake damage. Basic Appl Soc Psychol 23(2):109–121

McClure J, Sutton JM, Sibley C (2007) Listening to reporters or engineers: how instance-based messages about building design affect earthquake fatalism. J Appl Soc Psychol 37(9):1956–1973

McClure J, Wills C, Johnston D, Recker C (2011) How the 2010 Canterbury (Darfield) earthquake affected earthquake risk perception: comparing citizens inside and outside the earthquake region. Australas J Disaster Trauma Stud 2011–2:3–10

Mclvor D, Paton D, Johnston D (2009) Modelling community preparation for natural hazards: understanding hazard cognitions. J Pac Rim Psychol 3(2):39–46

Mileti DS, O'Brien PW (1992) Warnings during disaster: normalizing communicated risk. Soc Probl 39(1):40–57

Ministry of Civil Defence & Emergency Management (2006a) The way forward – strategic framework for the National CDEM Public Education Programme 2006–2015. Ministry of Civil Defence & Emergency Management, Wellington

Ministry of Civil Defence & Emergency Management (2006b) Colmar Brunton Research Ltd: campaign monitoring research 2006. Ministry of Civil Defence & Emergency Management, Wellington

Ministry of Civil Defence & Emergency Management (2007) Colmar Brunton Research Ltd: campaign monitoring research 2007. Ministry of Civil Defence & Emergency Management, Wellington

Ministry of Civil Defence & Emergency Management (2008a) Colmar Brunton Research Ltd: campaign monitoring research 2008. Ministry of Civil Defence & Emergency Management, Wellington

Ministry of Civil Defence & Emergency Management (2008b) National tsunami signage, technical standard [TS01/08]. Ministry of Civil Defence & Emergency Management, Wellington

Ministry of Civil Defence & Emergency Management (2008c) Tsunami evacuation zones, director's guideline for Civil Defence Emergency Management Groups [DGL 08/08]. Ministry of Civil Defence & Emergency Management, Wellington

Ministry of Civil Defence & Emergency Management (2009) Colmar Brunton Research Ltd: campaign monitoring research 2009. Ministry of Civil Defence & Emergency Management, Wellington

Ministry of Civil Defence & Emergency Management (2010) Colmar Brunton Research Ltd: campaign monitoring research 2010. Ministry of Civil Defence & Emergency Management, Wellington

Ministry of Civil Defence & Emergency Management (2011) Colmar Brunton Research Ltd: campaign monitoring research 2008. Ministry of Civil Defence & Emergency Management, Wellington
Ministry of Civil Defence & Emergency Management (2012) Colmar Brunton Research Ltd: campaign monitoring research 2008. Ministry of Civil Defence & Emergency Management, Wellington
Mulilis J-P, Duval TS (1995) Negative threat appeals and earthquake preparedness: a Person-relative-to-Event (PrE) model of coping with threat. J Appl Soc Psychol 25(15):1319–1339
Mulilis J-P, Duval T (1997) The PrE model of coping and tornado preparedness: moderating effects of responsibility. J Appl Soc Psychol 27(19):1750–1765
New Zealand Police (2012) List of deceased, as at 9 Feb 2012. New Zealand Police. http://www.police.govt.nz/list-deceased. Accessed 23 Feb 2012
Paton D (2005) Community resilience: integrating hazard management and community engagement. Paper presented at the proceedings of the international conference on engaging communities, Brisbane
Paton D (2006) Disaster resilience: integrating individual, community, institutional and environmental perspectives. In: Paton D, Johnston D (eds) Disaster resilience: an integrated approach. Charles C Thomas, Springfield, pp 305–318
Paton D (2008) Risk communication and natural hazard mitigation: how trust influences its effectiveness. Int J Glob Environ Issues 8(1):2–16
Paton D (2010) Adaptive capacity/resilience model: summary of PGSF research. Joint Centre for Disaster Research, Wellington
Paton D (2012) MCDEM Christchurch community resilience project report: part 1. Ministry of Civil Defence and Emergency Management, Wellington
Paton D, Johnston D (eds) (2006) Disaster resilience: an integrated approach. Charles C Thomas, Springfield
Paton D, McClure J (in preparation) Preparing for disaster: building household and community capacity. Charles C Thomas, Springfield
Paton D, Smith L, Johnston D (2000) Volcanic hazards: risk perception and preparedness. N Z J Psychol 29(2):84–88
Paton D, Smith L, Johnston D, Johnston M, Ronan K (2003) Developing a model to predict the adoption of natural hazard risk reduction and preparatory adjustments, EQC research project no. 01-479. Earthquake Commission, Wellington
Paton D, Smith L, Johnston DM (2005) When good intentions turn bad: promoting natural hazard preparedness. Aust J Emerg Manag 20(1):25–30
Paton D, McClure J, Bürgelt PT (2006) Natural hazard resilience: the role of individual and household preparedness. In: Paton D, Johnston D (eds) Disaster resilience: an integrated approach. Charles C Thomas, Springfield, pp 105–124
Perry RW, Lindell MK (2008) Volcanic risk perception and adjustment in a multi-hazard environment. J Volcanol Geotherm Res 172(3–4):170–178
Rip A (1988) Should social amplification of risk be counteracted? Risk Anal 8(2):193–197
Russell LA, Goltz JD, Bourque LB (1995) Preparedness and hazard mitigation actions before and after two earthquakes. Environ Behav 27(6):744–770
Rüstemli A, Karanci AN (1999) Correlates of earthquake cognitions and preparedness behavior in a victimized population. J Soc Psychol 139(1):91–101
Said AM, Ahmadun F-R, Mahmud AR, Abas F (2011) Community preparedness for tsunami disaster: a case study. Disaster Prev Manag 20(3):266–280
Showalter PS (1993) Prognostication of doom: an earthquake prediction's effect on four small communities. Int J Mass Emerg Disasters 11(3):279–292
Siegel JM, Shoaf KI, Afifi AA, Bourque LB (2003) Surviving two disasters: does reaction to the first predict response to the second? Environ Behav 35(5):637–654
Solberg C, Rossetto T, Joffe H (2010) The social psychology of seismic hazard adjustment: re-evaluating the international literature. Nat Hazards Earth Syst Sci 10(8):1663–1677

Spittal MJ, McClure J, Siegert RJ, Walkey FH (2005) Optimistic bias in relation to preparedness for earthquakes. Australas J Disaster Trauma Stud 2005–01:1–10

Spittal M, McClure J, Walkey F, Siegert R (2008) Psychological predictors of earthquake preparation. Environ Behav 40(6):798–817

Turner RH (1983) Waiting for disaster: changing reactions to earthquake forecasts in Southern California. Int J Mass Emerg Disasters 1(2):307–333

Turner RH, Nigg JM, Paz DH (1986) Waiting for disaster: earthquake watch in California. University of California Press, Los Angeles

Wallace LM, Reyners M, Cochrane U, Bannister S, Barnes PM, Berryman K et al (2009) Characterizing the seismogenic zone of a major plate boundary subduction thrust: Hikurangi margin, New Zealand. Geochem Geophys Geosyst 10(10):Q10006

Weinstein ND, Klein WM (1996) Unrealistic optimism: present and future. J Soc Clin Psychol 15(1):1–8

Wellington Region Civil Defence Emergency Management Group (2005) Wellington Region Civil Defence Emergency Management Group Plan. http://www.gw.govt.nz/assets/council-publications/cdem_plan_may2005_web.pdf. Accessed 28 Mar 2011

Whitney DJ, Lindell MK, Nguyen HHD (2004) Earthquake beliefs and adoption of seismic hazard adjustments. Risk Anal 24(1):87–102

Chapter 9
Perceptions of Climate Variability and Coping Strategies in Informal Settlements in Dhaka, Bangladesh

Huraera Jabeen and Cassidy Johnson

Abstract Many of the urban poor living in Dhaka city, Bangladesh are frequently exposed to flood hazards, as most of the informal settlements are located in highly flood-prone areas of the city. Based on findings from household surveys in two informal settlements in Dhaka, this chapter seeks to examine the relationship between people's perceptions of flood hazards and the actions they take to reduce their vulnerability. It also explores the latent drivers (psychological, environmental and political) behind these perceptions and behaviours. The findings conclude that risk perceptions and related preparedness actions are framed by the highly vulnerable context in which the urban poor live in. Although they are extremely anxious about their exposure to a range of hazards or any life stresses, their limited power to influence institutional-level affairs encourages undertaking more individual and household preparedness through a range of different activities. The chapter suggests that the experience of relatively small hazard events on a yearly basis fosters highly-tuned perceptions about hazard risk and high levels of worry, thereby making the urban poor more resilient to a range of hazards and life stresses.

9.1 Introduction

Dhaka, the capital city of Bangladesh, is considered to be a city at high risk from the impacts of climate change. It is located in a hazard-prone region with high urban population density and substantial rates of urban poverty. The city also has a very

H. Jabeen (✉)
Department of Architecture, BRAC University,
66 Mohakhali, 1212 Dhaka, Bangladesh
e-mail: huraera@bracu.ac.bd

C. Johnson
Development Planning Unit, Faculty of the Built Environment, University College London, 34 Tavistock Square, WC1H 9EZ London, UK
e-mail: cassidy.johnson@ucl.ac.uk

H. Joffe et al. (eds.), *Cities at Risk: Living with Perils in the 21st Century*, Advances in Natural and Technological Hazards Research 33,
DOI 10.1007/978-94-007-6184-1_9,

large proportion of the population living in informal settlements, which lack adequate infrastructure and whose residents usually have no legal tenure status. The informal settlements generate a context of intense social vulnerability, where socioeconomic circumstances are creating situations in which people are highly susceptible to the impact of hazards and hazardous conditions.

Under these conditions, how do the urban poor – living in informal settlements in Dhaka – perceive hazards? Many of the preparations people make to deal with hazards occur without assistance from the local government or other government entities. In the absence of such assistance, how do the urban poor respond to the climate-related hazards they face and how do they behave in anticipation of them? To explore these questions, this chapter examines the results of a research project undertaken between 2008 and 2010 that aimed to understand how the impacts of climate change in Bangladesh will affect the urban poor of Dhaka and to understand how people are coping with natural hazards and hazardous conditions arising from climate change. The research is specifically related to climatic hazards (climate variability, flooding, water-clogging and heat in particular), and draws on 55 semi-structured interviews with residents of two different informal settlements in Dhaka. The interviews sought to look at people's behavioural coping strategies in particular, and in this chapter we seek to elucidate the relationship between these coping strategies and people's perceptions of the hazards they face.

Firstly, this chapter offers a contextual basis for understanding climatic hazards and urbanisation in Dhaka. Secondly, it analyses the household responses collected to understand hazard perceptions and actions taken to reduce vulnerability. This is followed by a discussion of the latent drivers (psychological, environmental and political) behind these perceptions and behaviours. The chapter tries to shed light on how informal urbanisation, poverty and power relations impact on people's perceptions of hazards and the actions they take as a result.

9.2 Dhaka: Urbanisation and Natural Hazards

Dhaka is naturally prone to fluvial flooding. Its geographical location is in a delta, surrounded by rivers and water-bodies on all four sides. This is coupled with an insufficient citywide drainage system, which is unable to cope with intense rainfall. Thus, the city is also exposed to increased risks from pluvial flooding. Most literature on Dhaka describes the threats of flooding by citing examples of nine major floods over the last 55 years. A multitude of factors contributed to these events, including overflow of surrounding rivers, excessive untimely rainfall, drainage congestion, and inadequate pumping facilities (Alam and Rabbani 2007; Dewan et al. 2006; Stalenberg and Vrijling 2005). A map, based on land cover and elevation data for flood frequency and floodwater depth in greater Dhaka area, showed 19.2% of the area to be in a very high hazard zone, with higher exposure to floods in areas with lower elevation (Dewan et al. 2006).

Furthermore, in recent years differentials in temperatures during summer and winter have been making headlines in newspapers depicting the misery of the city-dwellers. The combined effect of severe heat with temperature nearing 40°C, repeated electricity load shedding, acute water crisis, shortage of gas supply, infestations of mosquitoes and the resultant near epidemic of diseases such as diarrhoea during summer are impacting on people's well-being (The Daily Star 2010). The extreme pre-monsoon heat also causes unpredicted, intense rainfall, resulting in havoc all over the city (The Daily Star 2009). During winter, temperature often drops to 8°C, which can acutely disrupt day-to-day life and even claim the lives of those unprepared to cope with such temperatures (The Daily Star 2011). These kinds of severe weather conditions and extreme climate events are not always recorded as disasters, but affect how the city-dwellers respond to the risks they face in their everyday life.

Although all city-dwellers are exposed to these climatic hazards and hazardous conditions, vulnerability varies between different groups living in the city. In Dhaka, as with many other cities, it is argued that the urban poor are most vulnerable to the impacts of climate change (Baker 2012; McGranahan et al. 2007; Pelling 2003; Satterthwaite et al. 2009). A census conducted in 2005 indicated that 5.4 million urban poor of the six major cities in Bangladesh lived in 'slums' situated on only 4% of the land area of those cities; among them 3.4 million – or 63% – lived in Dhaka alone. Moreover, between 1996 and 2005, not only did the total population living in these settlements more than double (from 1.5 to 3.4 million) in Dhaka; but also the proportion of urban poor increased from 20 to 37% of the total population (CUS, NIPORT and MEASURE Evaluation 2006). No official census has since been published to indicate the changing proportion of the city's total population, which is estimated to have reached more than 15 million in 2011, compared to the nine million when this study was conducted in 2005. Similarly, the size of the urban poor population is estimated to have increased.

As in any other city in a low- or middle-income country, the urban poor of Dhaka experience greater risks from natural hazards due to greater exposure (e.g. through living in makeshift housing on unsafe sites); lack of hazard-reducing infrastructure (e.g. drainage systems, roads allowing emergency vehicle access); less adaptive capacity (e.g. the ability to move to better quality housing or less dangerous sites); less state provision for assistance in the event of a disaster (e.g. state action may increase exposure to hazards by limiting access to safe sites for housing); less legal and financial protection (e.g. a lack of legal tenure for housing sites, lack of assets and insurance); deficient information, communication and knowledge (e.g. where to move and when); and absence of institutional and community organization (Cardona 2004; Dodman and Satterthwaite 2008). Hence, they suffer disproportionately from direct impacts of increased and untimely rainfall, flooding and cyclone events; increased heat and warm spells as well as indirect impacts of water-clogging and landslide; water shortage; heat islands; disruption of livelihoods and incomes; loss of assets and infrastructure; increased diseases and increased migration.

The climate risks for Dhaka are exacerbated by the high rate of urban population growth and urbanisation that add pressure to the existing, deficient infrastructure. As the capital city in a centralised economic, administrative and political system, Dhaka attracts between 300,000 and 400,000 migrants each year (Baker 2006) from all over the country due to better economic opportunities, education and health-care facilities and better living conditions. However, the administrative and institutional arrangements to direct urban development have limited capacity to cope with this surge. The city grew without any effective urban development plans where the rapidly growing demands from different sectors encouraged the planning authorities to create extended metropolitan regions. Hence, the infrastructure planning and development remained dispersed and inadequate relative to the demands of a growing population.

The high demand for land for development has resulted in high land prices, high density development and land speculation. Dhaka has experienced a property boom in recent decades without any concern for available infrastructure; moreover, some real-estate developers have encroached on the natural depressions and waterways beyond the jurisdictions of the city development authorities. The combined effect of unequal development and management of the utility services, and improper management of the natural resources and natural hazards have degraded the overall environment of the city (Rabbani et al. 2011). As a consequence of high demand and land speculation, the urban poor of Dhaka are left with no other choice but to live in the very areas in which they will be most vulnerable.

In addition, policies and institutional arrangements for adaptation planning and urban development are not favourable for the urban poor. Rosenzweig et al. (2011) argue that a central challenge for a city is not only to establish the links between climate change, adaptation and disaster risk reduction, but also with climate change policies. If the policies do not address the risks associated with climate change for different groups appropriately, they themselves can aggravate vulnerability. The two major policy initiatives – the 2005 National Adaptation Programme of Action (NAPA) and the 2008 Bangladesh Climate Change Strategy and Action Plan (BCCSAP) – remain deficient in their addressing of urban issues, especially vulnerabilities of the urban poor. Although the latter recognised that with rapid and unplanned urbanisation in Bangladesh the vulnerabilities of the urban poor will become an even more urgent and pressing problem, the strategy did not spell out any specific activities to address the problem (MoEF 2008).

Similarly, urban development strategies and plans in Dhaka conceive the urban poor as 'illegal' in the city and hence marginalise them in accessing housing and infrastructure. Most of the informal settlements are either squats in public land or constructed in private land by intermediaries who have implicit agreements with the land-owners. As a consequence, without any legal address they are commonly denied basic rights and entitlements, including the right to access water, sanitation, healthcare services, and education (Rashid 2009). Climate-induced hazards tend to increase in scale in the absence of access to housing, infrastructure and services.

9.3 The Case Study Areas

Fifty-five households in two different locations in Dhaka (Korail and Mohammadpur beribadh in Fig. 9.1) were interviewed using a qualitative, semi-structured questionnaire to identify their experiences of climate risks, hazards and coping strategies. The choice of these two study areas was based on the following variables: location in the context of the city, land and house-ownership patterns, housing and infrastructure conditions, livelihood opportunities, access to social safety nets and population patterns.

Similar characteristics also guided the choice of households within the communities. Households were selected randomly based on where they were located: near the water edge and susceptible to flooding, or on higher ground in high density areas with or without drainage facilities, susceptible to impacts of extreme heat and water-clogging. The households were a mixture of house-owners and renters living in the areas for more or less than 5 years in houses made out of different building materials of varying conditions. The dwellings selected ranged from structures made out of bamboo and corrugated iron sheets to bricks, and from permanent to temporary conditions. After many transect walks through the areas and interviewing different stakeholders, these characteristics were utilised to identify households in possible locations. Later some households were approached. Only the responses of the households who volunteered to respond to the semi-structured interviews were recorded.

Interviewers used semi-structured interviews with the respondents and asked about the basic household information, tenancy history, finance and assets, and made notes about the physical conditions of the room/dwelling. Through informal discussion respondents were asked about the impact of the weather and physical environment on their household and house. They were not asked directly about climate change, nor was climate change mentioned by the interviewer. They were also asked about their experiences and actions over the last years during disasters. The responses of the semi-structured interviews were compiled and analysed using the programmes Microsoft Office Access, Excel and SPSS. Notes of the comments, pictures and sketches provided the qualitative data of the research.

9.3.1 Korail

Korail, the first area of study, is the largest informal settlement in Dhaka. More than 18,000 households (Urban Partnership for Poverty Reduction Project 2010), or an estimated population of more than 100,000 (CUS, NIPORT and MEASURE Evaluation 2006), live in this squatter settlement developed in 90 acres of government land since the 1980s. The community live under constant threat of eviction due to land-ownership disputes and extremely high land price of the surrounding high-income residential and commercial areas. Lack of security of tenure and associated institutional arrangements also restrict the residents from getting

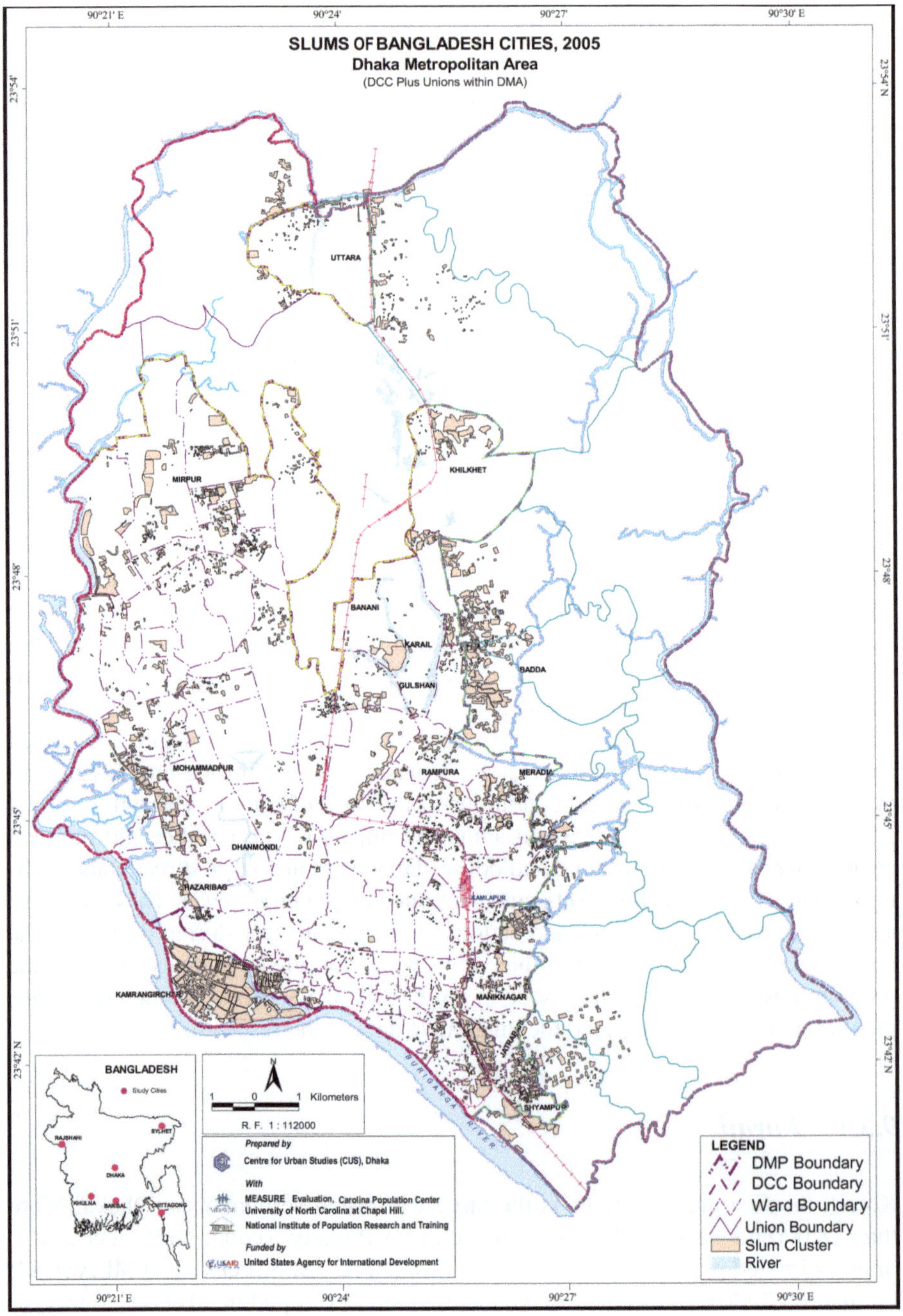

Fig. 9.1 Location of Korail and Mohammadpur beribadh in Dhaka city (*Source*: CUS, NIPORT and MEASURE Evaluation 2006)

Fig. 9.2 Aerial view of part of Korail with high-income residential area in the background (Photo credit: Huraera Jabeen)

approval for constructing housing or accessing infrastructure or services from city authorities.

These socio-political vulnerabilities exacerbate Korail residents' physical vulnerabilities. The area is surrounded by a water reservoir on two sides that often causes fluvial flooding; inadequate drainage facilities within the settlement cause water-clogging inside the houses and roads during rainfall. Similarly the high density developments with high heat-emitting materials are vulnerable to the effects of increased temperature. Nevertheless, high demand for housing and services has encouraged the residents to develop the settlement on their own where they share a house with multiple households in different rooms, with common use of services like toilets, washing areas and kitchens. To support these services an informal system of supply of services has been developed in the area by the residents (Fig. 9.2).

Access to affordable housing with services, along with access to livelihood opportunities for both women and men, has attracted a high concentration of low-income households in Korail. Most of the livelihood opportunities available are low-paid and labour-intensive. The women either work in the ready-made garments industries or as household help in the surrounding areas; whereas most men are engaged in pulling rickshaws or running small businesses as street vendors. Many households have home-based businesses within the area, where the female members participate to complement household incomes alongside taking care of household activities.

Fig. 9.3 One of the liner houses in Mohammadpur beribadh (Photo credit: Huraera Jabeen)

9.3.2 Mohammadpur Beribadh

The second study area in Mohammadpur beribadh is one of the largest clusters of informal settlements located inside the flood protection embankment along the western fringe of the city. The settlements were developed on private land by either land-owners or their intermediaries. Although these informal settlements are popular with first migrants to the city due to affordable housing and access to livelihood opportunities as unskilled labours, the residents often face threats of eviction when the land-owners decide to lease the land to others or develop for other purposes (for example, constructing housing facilities targeted at higher-income groups).

The key sources of physical vulnerability in these settlements are pluvial flooding and water-clogging. The embankment has significantly reduced risks from fluvial flooding; nevertheless, the settlement remains water-clogged after any rainfall when the run-off from the surrounding areas accumulates on the low-lying area inside the embankment. With few pumping stations in the area it takes a considerable amount of time to drain out the stagnant water. In addition, the settlement is very close to the tannery industries, whose industrial wastes pollute the nearby water-bodies. When the run-off mixes with this polluted water and remains stagnant, the situation further deteriorates (Fig. 9.3).

The land-owners are reluctant to invest in housing and infrastructure development for the high cost of developing in low-lying land near the tannery industries.

The living conditions of the settlements are very poor compared to Korail, with houses constructed of bamboo mats or plastic sheets on stilts, and mostly with hanging latrines. Most of these latrines dump waste into the stagnant water below, on which the settlement developed. Nevertheless, private land-ownership in Mohammadpur beribadh created some opportunities to access services such as electricity and water from the government authorities.

Most women of these settlements work as household help in the adjoining middle to higher-middle income, high-density residential areas or day labour, whereas most men are employed in day labour. However, their wages are less than the wages received by those living in Korail. As a consequence, the renters are usually of a transient character, moving from one settlement to another depending on the availability of job opportunities and cheaper rent.

9.4 Findings

9.4.1 Perception and Impacts of Natural Hazards and Climate Change

The household interviews aimed to understand how the residents perceived climate change and whether changes in the weather patterns were apparent to them or not. Accordingly, questions were asked about what respondents thought had changed over the last few years in the climate/weather pattern and how they would describe these changes. The environmental change variables in Table 9.1 were developed from the responses.

In both of the study areas, respondents reported increased heat as the most prominent change that has occurred over recent years (86.7% in Korail and 92% in Mohammadpur beribadh). They associated increased heat with fatigue and discomfort, in addition to increased possibilities of water shortages that result in a greater risk of water-borne diseases, such as typhoid and diarrhoea. With regards to rainfall, respondents' opinions about changes varied between households and communities. In many instances, the older respondents asserted a decrease in amount of rainfall from their childhood experiences. Twelve out of thirty respondents in Korail and 13 out of 25 respondents in Mohammadpur beribadh perceived a change in rainfall patterns rather than rainfall amount. In addition, the respondents in Mohammadpur beribadh identified major risks to be the flash flooding associated with changed rainfall patterns and protracted flooding due to lack of drainage; on average 52% of the respondents reported these risks.

In general, respondents' descriptions of changes in climate are related to climate variability, severe weather conditions and extreme events that they experience on a yearly or frequent basis. During the interviews the respondents expanded their responses by describing and explaining their everyday sufferings caused by increased heat, water-clogging or pluvial flooding. Long-term changes in climate such as global warming or sea-level rise did not bear any significance to the respondents.

Table 9.1 Responses about environmental changes

	Korail (*N*=30)		Mohammadpur beribadh (*N*=25)	
Environmental changes	Number	Percentage	Number	Percentage
Increased heat	26	86.7	23	92.0
Increased rainfall in shorter time period	13	43.3	11	44.0
Rainfall at different times than normally expected	12	40.0	13	52.0
Flood waters staying for a longer period	11	36.7	13	52.0
Less rainfall	10	33.3	18	72.0
Flash flooding	8	26.7	13	52.0
Air pollution	8	26.7	11	44.0

Frequent, small-scale events dominated their perception of hazards rather than infrequent hazards that are recorded as disasters by the government authorities.

The impacts of these changes were also probed in open-ended questions to explore the perception of risks. The responses reflected respondents' worries and anxieties about the severity of the hazards and their effects on their various physical, financial and human assets.

In Korail, respondents reported damage of home (86%) and damage of infrastructure (76%) as the most dominant impacts of any natural hazards. In Mohammadpur beribadh, damage to access facilities was the most dominant impact (80%) with damage of home (76%) second most important. Most of the respondents reported some form of loss of household assets during any hazardous events. However, only 20% in both areas reported displacement as an impact; people tend not to move from their houses during disasters due either to an inability to move or to fear of losing possession of their house and other assets.

The physical vulnerabilities from any natural hazards translate into financial and human risks. A considerable proportion of the respondents reported loss of working days as a direct impact of natural hazards (60% in Korail and 72% in Mohammadpur beribadh). It is important to point out that loss of livelihood and business capital was a more prevalent response in Mohammadpur beribadh, where more people work as day labourers or street vendors, capitalising on human assets to earn.

In both areas, respondents discussed increased occurrence of diseases and the spread of new diseases, such as dengue, with reference to how they impact upon the well-being of the earning household members. Moreover, they related these impacts to increased health expenditure: when earning members suffered from these diseases they lost working days and hours, and thus earnings; when other members – especially children – suffered, the unexpected medical expenses also added pressure to their limited household earnings. On average, more than 70% of respondents reported suffering from such indirect impacts.

Table 9.2 Impacts of natural hazards

Impacts	Korail ($N=30$)		Mohammadpur beribadh ($N=25$)	
	Number	Percentage	Number	Percentage
Damage of home	26	86.7	19	76.0
Damage of infrastructure	23	76.7	14	56.0
Water-clogging	22	73.3	18	72.0
Increased waterborne disease	22	73.3	17	68.0
Damage of households assets	20	66.7	13	52.0
Loss of working days	19	63.3	18	72.0
Increased health related expenditure	19	63.3	19	76.0
Damage of access facilities	18	60.0	20	80.0
Loss of livelihood/business capital	17	56.7	15	60.0
New disease (malaria/dengue/ tuberculosis)	12	40.0	4	16.0
Increased respiratory disease	6	20.0	11	44.0
Displacement	6	20.0	5	20.0

The diverse responses related to the perceptions of changes and impacts of natural hazards (shown in Table 9.2) clearly reflect worries about uncertainty experienced by the respondents. They ranged from physical and financial to social vulnerabilities. However, these perceptions seemed to guide different actions taken to reduce the risks in different phases of natural hazards and hazardous conditions. The following section discusses those actions reported by the respondents, based on their resources and capabilities.

9.4.2 Actions Taken to Reduce Physical Vulnerabilities

It is argued that the perception and impacts of changes are formulated through how people experience their vulnerabilities; thus, both perception and experience create the basis for decision making and for taking actions to reduce vulnerability (Vedwan and Rhoades 2001). This study showed that the perceptions and actual impacts of the natural hazards motivated residents to take some actions to address their vulnerability. When queried about actions taken before, during and after any natural hazards, the respondents reported activities that were taken both at the household and community levels. It seemed that all the respondents accepted that they did not have any control over the changes in environmental variables, defining them as natural changes; they therefore took some control of their actions at the household level to reduce the impacts, while they had limited control over the social and political forces that exacerbated the climate vulnerabilities at the community level. Consequently, the responses varied in the two study areas based on these socio-economic and political contexts.

Table 9.3 Actions taken before and after any natural disasters

	Korail (*N*=30)		Mohammadpur beribadh (*N*=25)	
Actions taken	Number	Percentage	Number	Percentage
Before any natural disasters				
Increase height of furniture	16	53.3	16	64
Make barriers at the door	13	43.3	3	12
Make higher plinth	9	30.0	5	20
Make higher storage facilities	9	30.0	18	72
Store food and water	5	16.7	6	24
Change building materials	5	16.7	9	36
Construct drainage	3	10.0	2	8
Improve drainage system	2	6.7	0	0
Move family to safer areas	1	3.3	4	16
Remove/relocate service lines	0	0.0	1	4
After any natural disasters				
Rebuild structure	18	60.0	7	28
Increase height of plinth/sill level/doors/windows	11	36.7	7	28
Take loan for rebuilding	9	30.0	5	20
Help community members to rebuild	7	23.3	2	8
Change building materials	7	23.3	2	8
Link drainage to main system	5	16.7	1	4
Share resources with neighbours	4	13.3	2	8
Move to new location	1	3.3	3	12

At the household level, respondents took various actions before and/or after any natural disasters to reduce the impact of water-clogging or flooding, for example making higher storage facilities attached to partitions or above false ceilings to place their belongings above water-levels, raising furniture to reduce loss from water-clogging, using more resilient building materials, building the house on higher plinths, and making barriers at doors to stop water coming inside the rooms. The activities taken at the community level before any events of water-clogging were limited to constructing or improving drainage facilities. Another important aspect to highlight from these findings is that, when asked about the information contributing to decisions about these actions, especially at the household level, most respondents stated that they learned from their own and their neighbours' experiences.

However, as is evident from comparing the responses in the two studied areas in Table 9.3, the priority and frequency of taking these actions varied between respondents. For example, in Mohammadpur beribadh, where flash flooding and water-clogging are frequent, making higher storage (as described above) is the most practiced action; whereas respondents in Korail described making higher cement plinths to reduce the impact of water-clogging. The responses may be related to the housing and settlement conditions, security of tenure and economic conditions.

In Korail, people squatting in government land with better economic opportunities were able to construct houses with CI sheets that were more water resilient. In contrast, the transitory houses in Mohammadpur beribadh, built on stilts of bamboo mat and occupied by the extremely poor households, were exposed to water-seepage through partitions and floorings. They had limited resources with which to increase their house's resilience.

Usually the preventive activities, for example, increasing plinth height to reduce impact of water-clogging or making openings for better ventilations, were guided by the respondents' experiences of the impact of previous hazards. Hence, 60% of the respondents in Korail reported rebuilding houses after extreme events, as they caused some form of destruction to the floors, structural members or partitions. Accordingly, they changed their choice of building materials, increased the height of the plinth and made windows for ventilation in the rooms. Some respondents (30%) took loans for rebuilding houses whereas others (23%) reported receiving physical help from community members. All of these preventive actions depended on whether or not the household members had suffered during the previous events, and whether or not they felt the changes would improve their home's resilience.

The respondents in Mohammadpur beribadh reported having less control over their living conditions, and accordingly reported having taken fewer actions towards repairing or rebuilding in the aftermath of disasters. This may be because the economic hardships of these households were harsher than those in Korail. In addition, the transient nature of the renters and land tenure-ship seems likely to have reduced the inclination of house-owners to invest in housing improvements. Concomitant with any investment in a house is higher rent. Extremely poor households – such as those living in Mohammadpur beribadh – cannot afford such increases in expenditure. Therefore, short-term renters consider moving to a new location or stay with worse living conditions. In Korail, the residents tend to stay in the same area, if not in the same house, and so repairing and rebuilding houses is more common both for the house-owners and renters.

The vulnerability of the respondents in Mohammadpur beribadh is further evident in the types of actions taken during any disaster in comparison to respondents in Korail (Table 9.4). Those in Mohammadpur beribadh suffered most from food shortage; they used furniture, for example beds, as the only refuge from water-clogging while using movable ovens to cook food, sometimes even on beds. Fifty-two percent of Mohammadpur beribadh respondents reported building stilts inside their rooms to cope with increased water levels during flooding. In comparison, although the respondents in Korail also used furniture and movable ovens as means of coping, they seemed to have comparatively more control over their predicament, as demonstrated by their clearing of drainage, making outlets to ensure drainage of water, developing alternate means of access, and sharing services with their unaffected neighbours. However, in both areas, most respondents reported borrowing money to tackle hardships, indicating their anxiety in regard to coping with the uncertain future and their preparedness. The adoption of different financial strategies and efforts to capitalise upon human assets are described in the following sections.

Table 9.4 Actions taken during any natural disasters

	Korail (N=30)		Mohammadpur beribadh (N=25)	
Actions	Number	Percentage	Number	Percentage
Sleep on the furniture above flood level	14	46.7	18	72.0
Use movable cooker	11	36.7	17	68.0
Share services of unaffected neighbours	9	30.0	4	16.0
Suffer from food shortage	9	30.0	15	60.0
Borrow money to tackle hardship	8	26.7	7	28.0
Outlet at houses for easy flow of water	6	20.0	1	4.0
Move family to safer areas	6	20.0	9	36.0
Make barriers at the door	6	20.0	4	16.0
Move assets	5	16.7	7	28.0
Clear drainage	5	16.7	2	8.0
Food sharing	5	16.7	8	32.0
Organize community efforts	5	16.7	4	16.0
Build stilt inside the house	5	16.7	13	52.0
Build/arrange emergency shelter	4	13.3	1	4.0
Develop alternate means of access	3	10.0	2	8.0

There were wide variations between individual as well as combined household earnings within the respondents in Korail and Mohammadpur beribadh. In Korail more than 60% of the studied households had a combined monthly earning of more than BDT 5000 (USD 71.5); which the census of settlements of six cities identified as average income (CUS, NIPORT and MEASURE Evaluation 2006). In comparison, the respondents from Mohammadpur beribadh were more economically poor than those of Korail. Fifteen out of twenty five households (60%) had combined earnings of less than this average household income for supporting on average four members. In both the areas, dependency of non-earning members on earning members was quite high; 59.5% in Korail and 65% in Mohammadpur beribadh. These non-earning members include children, women and unemployed members of the households.

The employment patterns among the respondents were also mixed; 20% of the study population were self-employed, mostly working as rickshaw pullers, street vendors, or running small businesses. Since these jobs depend on the daily availability of work, the earnings they provide are vulnerable to natural hazards or hazardous conditions. Twenty-three percent of the respondents in Korail and 19% in Mohammadpur beribadh worked as service providers and were employed in private sectors in occupations such as household-helper, security guard and garment workers. Earnings from these professions are similarly vulnerable since unsecured monthly wages are affected by the loss of working days and hours.

Despite the economic poverty in both the areas, over 50% of households reported that they saved regularly through both formal and informal organizations, for example, life insurance companies, NGO-based saving groups and co-operatives. Depending on income and household members, the savings comprised an average of 10% of the household income in Korail and 6% in Mohammadpur (Fig. 9.4). When discussing

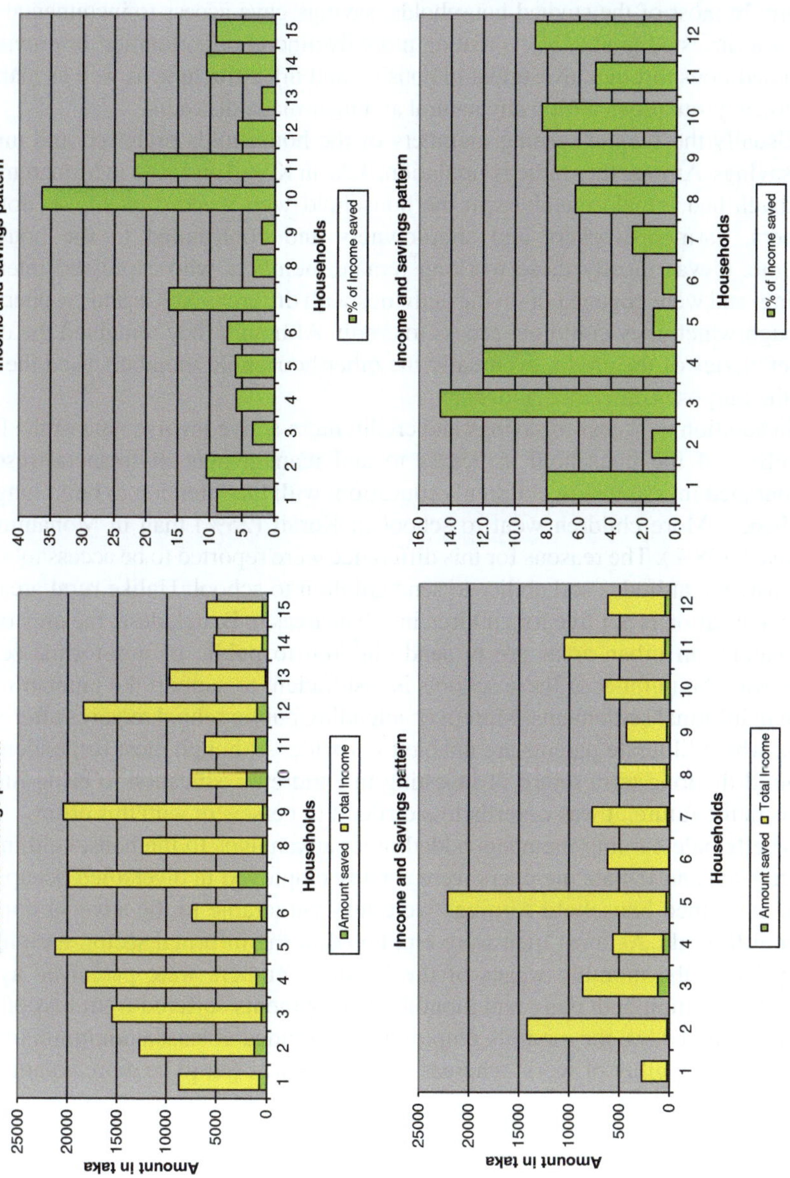

Fig. 9.4 Income and savings pattern of the studied households

the motivation to save, most respondents mentioned their worry of uncertainty for the future and savings were perceived as one of the key assets that allowed a household to guard against such uncertainty. Moreover, the savings generated an aspiration for a better life as they suggested examples of how saving can change things for better in future. In most of the studied households, savings gave access to incremental credit that was invested gradually in creating more livelihood opportunities, improving living conditions through investment in housing and infrastructure, as well as managing emergency situations during any natural and man-made disasters.

Usually the female earning members of the households preferred and initiated the savings. Among the studied population, 32% in Korail and 40% in Mohammadpur beribadh had female members in the household who worked mostly as domestic helpers, garment-workers and shop-owners and contributed to the household income. It was mostly these working female members who expressed interest in saving, and were engaged in saving activities with different NGOs and co-operatives, through which they could get access to credit. Although they remained the official beneficiaries of the credit, eventually the other household members used the credit for the purposes described above.

In addition to access to savings and credit, more active involvement of the female members of the household in access to and management of financial resources encouraged investment in children's education, with the intention to build long-term resilience. More children went to school in Korail (25%) than in Mohammadpur beribadh (18%). The reasons for this difference were reported to be access to schools and parents' attitudes and ability to send children to school. Unlike rural areas, primary education is not free for children in urban areas in Bangladesh, the only options for parents in urban areas are to send children to public or non-formal schools. However, the number of these schools is insufficient to support the number of children in informal settlements. Moreover, attending public school requires after-school help, which illiterate parents are unable to provide. Although most respondents recognised the long-term return of investing in children's education to bring financial gains in the future, it was nevertheless difficult to persevere with this effort.

The female earning members added other advantages to the household income. Since male and female members were usually employed in diversified occupations, their combined household earnings were less vulnerable to the adverse impact of natural hazards. As most men were employed in the informal sector, earning on a daily basis, the monthly wages of the female members were perceived as more secure. Although both daily and monthly employments suffered from loss of working hours and days, the monthly employment provided at least a minimum of assurance for availability of work, whereas daily work was felt to be more volatile.

9.4.3 Capitalising on Human Assets

As expressed by the respondents, one of the key factors contributing to reducing physical, financial and social vulnerabilities of households was access to human assets, in the form of social networks, knowledge, skills and education. Respondents reported using their social network when settling down in a new city, tackling hardship during disasters, and in long-term physical asset accumulation. Relatives and friends from original locations helped in securing living accommodation and employment opportunities and later became useful in accessing better housing, infrastructure, savings and credit. Fifty-six percent of respondents in Korail and 64% in Mohammadpur beribadh had such relatives or friends in the area in which they were living, and 76% of respondents of Korail had relatives and friends in the city. This social network did not always provide financial benefits, but they were reported to be valuable social capital in difficult times.

However, in the absence of such acquaintances, association with professional groups (for example, those employed in similar jobs) became significant in making a better living in the city. In Korail, 46% reported using their professional affiliations to access and improve their livelihood opportunities. If a new migrant wanted to work as a rickshaw-puller, or if someone wanted to change his/her job in a garment factory, they usually depended upon an introduction from someone who was already working in the desired profession. In such circumstances the social and professional networks proved to be beneficial.

Similarly in Mohammadpur beribadh, affiliation to political groups gave the residents some form of assurance against social and political vulnerabilities. The house-owners who work as guardians for the renters often encouraged them to participate in different political activities in return for small economic benefits. Moreover, the participations in political activities also gained some assurances of getting other assistance, for example food relief after flooding. The political groups play a significant role in determining the recipients of any government assistance; thus maintaining a good relationship with the political groups benefited the residents in Mohammadpur.

9.5 Discussion: Latent Drivers of People's Perceptions and Behaviours

9.5.1 Manifestations of Fatalism, Power and Responsibility for Taking Actions

With the exception of a few extreme flooding events in Dhaka (in 1988, 1998 and 2004), the hazards faced by households in Korail and Mohammadpur beribadh are relatively common occurrences, with water-clogging, pluvial flooding and extremes of heat occurring in relatively similar forms every year, albeit with some fluctuation in intensity. Fatalism, the belief that the destructive effects of a hazard are inevitable, does not appear to be prevalent throughout the sample population insofar as people do take actions to prepare for the event and to lessen its impact on the household. These preparations include both activities directly related to flood and water-clogging impacts, such as changes to the physical structure of the house, and also preparedness for a range of hazards, which includes investing in savings and nurturing social networks.

Fatalism is related to the degree to which people feel they have power to control the impact of a hazard (Paton and Johnston 2006). There are generally two different aspects that determine whether people have fatalistic attitudes toward hazards, or whether they will take action. People who have an internal locus of control believe that circumstances reflect their own actions and that they are therefore in control of these circumstances. These people will exert more control over their circumstances, trying to mitigate the impact of the hazard. People who have an external locus of control believe that their circumstances reflect societal forces and chance factors such as fate: these people will be less likely to mitigate the hazard on their own, instead believing that the authorities or others will be responsible (Paton and Johnston 2006).

Thus, it seems important to reflect upon what the respondents in the case studies have the power to control or to change. As described above, the people living in Korail and Mohammadpur beribadh are squatters on land that they do not own. They are generally very poor people with no welfare or livelihood security, except for that which they provide for themselves. Furthermore, squatting on public land and squatting on private land are different in terms of how people see their locus of control and reflect on people's action. Squatters on public land, such as those in Korail, have greater security of tenure since the land is essentially owned by the people (even if they regularly face the threat of eviction). We can infer from the responses of the interviewees that for squatters, trust in state land-owners is greater than trust in private land-owners.

It is quite commonly understood in the development literature that security of tenure is directly correlated with investments in housing. This is reflected in the greater levels of actions taken by residents of Korail in preparation for hazards and in reconstruction afterwards, compared with residents of Mohammadpur beribadh. In Korail, it was more frequent for people to make greater physical changes to their dwellings, such as changing the plinth height, improving the dwelling and taking out loans by which to do so. In Mohammadpur beribadh, the land is privately owned and the people living there expressed that they therefore do not feel in control of their circumstances as they could be forced to leave at any moment. This was reflected in the survey, as people were less likely in Mohammadpur beribadh to engage in hazard preparations and reconstruction, except for very temporary measures such as moving storage higher and raising furniture heights. People were also more likely to move in the event of a flood.

There are a few more general points related to fatalism in this context that are worth mentioning, even though they were not explored in the household interviews. One aspect is related to fate and poverty. Many of the risks that people in Korail and Mohammadpur beribadh faced are related to conditions of poverty that place people in vulnerable economic and social situations. It is interesting to question whether the poor see poverty as their fate, as an outcome of the prevailing economic system, or as within their control. One often hears in the informal settlements in Dhaka an echo of a fatalistic statement: 'we are poor, what can we do?' Furthermore, some expressed in the interviews that since they are economically poor it restricts their political and social power to change their circumstances.

In the urban context, the conditions of poverty and lack of control over their fate become even more acute. The poor are not accepted in urban life – if they were poor and lived in a rural area, they would have access to some food and water produced by themselves, whereas in urban areas this is not the case. People live in a cash-based economy in urban areas, where they need to pay for food, for water, for transport, and for access to all basic needs. Thus, poverty means something different in urban places; people can be much more economically and socially vulnerable in cities because of this monetary aspect of poverty and the external forces that maintain these conditions.

Another aspect worth mentioning is the religious aspect of fatalism, or the conviction that God is ultimately responsible for taking care of people – the ultimate external locus of control. In Bangladesh, if someone is unable to do something or something is beyond their capability, they often say, 'Inshallah, Allah dekhben', meaning hopefully God will take care. Nevertheless, responses to hazard risk are not wholly fatalistic, as people are undertaking preparations to cope with the effects of hazards and not simply leaving their fate up to God.

9.5.2 Frequent Flood Events, Community Knowledge: Implications for Worry, Anxiety and Preparedness

Worry related to flooding has been studied by several researchers. Paton and Johnston (2006) observe that unless a person perceives a threat or a risk it is unlikely they will be motivated to deal with it. With regard to disaster preparation, people interpret information in a context defined by their experiences, beliefs and conceptions. They also discuss how worry and anxiety is related to trust, which can be increased if people feel confident about the information source. According to Raaijmakers et al. (2008), the amount of worry about flood hazards is dependent on the expected severity of the hazard and its effects, such as housing or economic damage, health impacts and disruption to life. Messner and Meyer (2005) discuss how

> these judgments vary between individuals due to different levels of information and uncertainty, due to different intuitive behaviour, and also due to specific power constellations and positions of interest. As a consequence, the individuals of a community may assess the risk of being flooded very differently, because they do not have the same information about the probability of flood hazard events in their region, about flood mitigation measures and their effectiveness, and they perhaps have a different historical background regarding the experience of living in a floodplain and of being flooded (p. 7).

Knowledge about the severity and impacts of flooding in Korail and Mohammadpur beribadh is based solely on people's personal experiences from past events and on information from their neighbours and social networks. There is no information from the authorities about the potential for flooding, water-clogging or any other hazard. The worry that is generated exists because people in Korail and Mohammadpur beribadh have knowledge about the risk of flooding, the potential severity of the event and the possibility of other emergencies that they will have to deal with given very limited resources and limited support from the government.

Preparedness for the hazard events is based on this individual and community-level knowledge and, as observed, people undertake different approaches to addressing flood risk. Community networks are sharing knowledge about how to cope with these environmental issues and learning from each other through informal means. As mentioned, some respond to this worry by making physical changes to their dwelling, for example raising the plinth height and improving the drainage nearby. People are also engaged in saving: even if they have very little household income, households are saving on average 10% of their income to use in emergency situations. Further, people are developing strong social networks, which give them the sense that they will be supported by others in times of crisis. It is worth noting that people are not only preparing for floods or natural hazards, but savings and social networks represent preparations for a wide range of hazards or life-events that may affect them.

9.6 Conclusions

Understanding the perceptions of risk in the informal settlements of Dhaka has brought to light a number of key issues that can inform further research on risk perception. Firstly, risk perceptions and related preparedness actions of the very poor urban dwellers in Dhaka are framed by the highly vulnerable context in which they live. The respondents in this research live on land with varying degrees of insecurity of tenure, living with the constant threat of eviction and most are working informally for daily wages that are highly susceptible to any shocks or stresses. They do not have access to any formal welfare system or assistance from the authorities if they are affected by a disaster (although they may be assisted by NGOs to some degree for basic needs during a large disaster event). This urban context also means that people are living in a largely cash-based economy, where they need to pay for basic services, water, food and often transport to get to places of work. As pointed out earlier, this is different to the rural context, where it can be assumed that most people have some access to food, water and land no matter what their circumstances are.

Given all of these factors it is not surprising that these people are highly anxious about their exposure to a range of hazards or any life stresses (eviction, family illness, death in family, loss of income). They have little or no power to influence government-level hazard mitigation and exhibit little trust in land owners, and therefore all the responsibilities for preparedness measures fall to them. This seems to increase the level of worry, but also seems to encourage individual and household preparedness.

The research shows that individual and household preparedness include a range of different activities such as physical changes to their house, use of social networks and household savings and asset accumulation. With regards to social networks, savings and asset accumulation, people do not distinguish between hazards and other life stresses, but instead prepare for a range of possible negative events. Furthermore, the extent to which they perceive their tenure to be secure influences the type of preparedness measures they take: those with perceptions of higher tenure security are more likely to make substantial physical changes to their dwelling and invest in community-level infrastructure, whereas those who perceive their security of tenure to be weak have a greater tendency to leave the neighbourhoods in the event of a disaster or only make temporary provisions for the flooding.

The second key issue raised by this research is that the frequency of hazard events to which the respondents are subject gives them awareness about hazard risks. Pluvial flooding and water-clogging is happening on a yearly basis and so people's perceptions are based on actual experience of the events. However, the lack of communication about other hazard risks could perhaps mean that people are not aware of the likelihood of earthquakes, for example. Dhaka has a very high risk of earthquakes, although no large event has happened there in recent years, so it is likely people do not perceive this as a great risk (this is not something explored in our interviews).

An interesting point, or perhaps theory, is that these many small events that people are facing on a yearly basis foster highly-tuned perceptions about hazard risk and high levels of worry, thereby making them more resilient to a range of hazards and life stresses.

References

Alam M, Rabbani G (2007) Vulnerabilities and responses to climate change for Dhaka. Environ Urban 19(1):81–97

Baker JL (2006) Dhaka: improving living conditions for the urban poor. World Bank, Washington, DC

Baker JL (2012) Climate change, disaster risk, and the urban poor: cities building resilience for a changing world. World Bank, Washington, DC

Cardona O (2004) The need for rethinking the concept of vulnerability and risk from a holistic perspective. In: Bankoff G, Hilhorst D, Frerks G (eds) Mapping vulnerability: disasters, development, and people. Earthscan, London, pp 37–51

Centre for Urban Studies (CUS), National Institute of Population Research and Training (NIPORT), MEASURE Evaluation (2006) Slums of urban Bangladesh: mapping and census, 2005. www.cpc.unc.edu/measure/publications/tr-06-35. Accessed 12 Sept 2012

Dewan AM, Kumamoto T, Nishigaki M (2006) Flood hazard delineation in Greater Dhaka, Bangladesh using an integrated gis and remote sensing approach. Geocarto Int 21(2):33–38

Dodman D, Satterthwaite D (2008) Institutional capacity, climate change adaptation and the urban poor. IDS Bull 39(4):67–74

McGranahan G, Balk D, Anderson B (2007) The rising tide: assessing the risks of climate change and human settlements in low elevation coastal zones. Environ Urban 19(1):17–37

Messner F, Meyer V (2005) Flood damage, vulnerability and risk perception – challenges for flood damage research. UFZ discussion papers, 13/2005. www.ufz.de/export/data/1/26222_Disk_Papiere_2005_13.pdf. Accessed 15 June 2012

MoEF (2008) Bangladesh climate change strategy and action plan 2008. Ministry of Environment and Forests, Government of the People's Republic of Bangladesh, Dhaka

Paton D, Johnston DM (eds) (2006) Disaster resilience: an integrated approach. Charles C Thomas, Springfield

Pelling M (2003) The vulnerability of cities: natural disasters and social resilience. Earthscan, London

Raaijmakers R, Krywkow J, Veen A (2008) Flood risk perceptions and spatial multi-criteria analysis: an exploratory research for hazard mitigation. Nat Hazards 46(3):307–322

Rabbani G, Rahman AA, Islam N (2011) Climate change implications for Dhaka City: a need for immediate measures to reduce vulnerability. In: Otto-Zimmermann K (ed) Resilient cities: cities and adaptation to climate change: proceedings of the global forum 2010. Springer, Dordrecht, pp 531–541

Rashid SF (2009) Strategies to reduce exclusion among populations living in urban slum settlements in Bangladesh. J Health Popul Nutr 27(4):574–586

Rosenzweig C, Solecki WD, Hammer SA, Mehrotra S (2011) Urban climate change in context. In: Rosenzweig C, Solecki WD, Hammer SA, Mehrotra S (eds) Climate change and cities: first assessment report of the urban climate change research network. Cambridge University Press, Cambridge, pp 3–11

Satterthwaite D, Huq S, Reid H, Pelling M, Lankao PR (2009) Adapting to climate change in urban areas: the possibilities and constraints in low- and middle-income nations. In: Bicknell J, Dodman D, Satterthwaite D (eds) Adapting cities to climate change: understanding and addressing the development challenges. Earthscan, London, pp 3–50

Stalenberg B, Vrijling HAN (2005) The battle of Tokyo and Dhaka against floods. Built Environ 35(4):471–491

The Daily Star (2009) Downpour halts life. http://www.thedailystar.net/newDesign/news-details.php?nid=99057. Accessed 15 June 2012

The Daily Star (2010) Hit by heat: blistering summer makes Dhaka life unbearable. http://www.thedailystar.net/newDesign/news-details.php?nid=138218. Accessed 15 June 2012

The Daily Star (2011) Cold wave grips country: claims 3 more lives, fog disrupts flights, ferry movement. http://www.thedailystar.net/newDesign/news-details.php?nid=214615. Accessed 15 June 2012

Urban Partnership for Poverty Reduction Project (2010) Community household survey. Unpublished data

Vedwan N, Rhoades RE (2001) Climate change in the Western Himalayas of India: a study of local perception and response. Clim Res 19(2):109–117

Chapter 10
Risk Perception, Public Education and Disaster Risk Management

John Twigg

Abstract This concluding chapter discusses the relevance of the different ideas about hazard risk perception, written about in the rest of this book, to the practice of disaster risk management, particularly with regard to current moves to encourage community-based resilience. It identifies the diversity of views about how risk is perceived and how to study risk perception, noting the importance of socio-cultural factors in understandings of risk. It questions the relevance of much of the risk perception research carried out in high-income countries to the situations of poor and vulnerable people in low-income countries, who in addition to living with natural hazards have to manage many other threats to their daily well-being and livelihoods. The implications of these discussions for public education about risk are examined, as are related issues of responsibility, authority and trust.

10.1 Risk Perception and the Disaster Management Agenda

Risk perception is a key element of individual and collective disaster risk management. By increasing public understanding of hazard risks, disaster planners and managers seek to stimulate communities and individuals to take appropriate protective actions before and during crises (Twigg 2004, pp. 165–195).

The importance of this kind of activity is emphasised in the UN's Hyogo Framework for Action (HFA) 2005–2015, the international community's strategy for disaster reduction. The third of the HFA's five priorities for action is to "use knowledge, innovation and education to build a culture of safety and resilience at all levels", working on the principle that disasters "can be substantially reduced if people

J. Twigg (✉)
Department of Civil, Environmental and Geomatic Engineering,
University College London, Gower Street, London WC1E 6BT, UK
e-mail: j.twigg@ucl.ac.uk

H. Joffe et al. (eds.), *Cities at Risk: Living with Perils in the 21st Century*,
Advances in Natural and Technological Hazards Research 33,
DOI 10.1007/978-94-007-6184-1_10,

are well informed and motivated towards a culture of disaster prevention and resilience" (UNISDR 2005). Creation of a culture of safety requires widespread popular understanding of the significance of risks and subsequent willingness to take action to reduce them.

Hand in hand with this goes what the HFA calls a "more pro-active approach to informing, motivating and involving people in all aspects of disaster risk reduction in their own local communities" (UNISDR 2005). Community-based disaster risk management was originally advocated as a means of community mobilisation and empowerment where governments and officialdom were unable to act or unresponsive to local needs (Maskrey 1989). Today, it is written routinely into the policies and strategies of national governments and international agencies and justified on grounds of efficiency as well as accountability. The UK is one of many countries placing much greater emphasis on promoting community engagement in local-level actions to increase resilience to disasters (UK Cabinet Office 2011). Resilience, as Johnston et al. show, is an attractive concept in disaster planning because it can be applied to individuals, communities and institutions as well as the interactions between them.

10.2 Studying Risk Perception: Theories and Realities

For an ordinary individual member of the public, making judgements about risk is not simple. One must consider the nature of the hazards, the potential occurrence and impact of hazard events, the range of alternative actions and the consequences of each possible alternative. Appraisal of these events and alternative actions is not separate but interdependent. People are rarely aware of all the alternatives available to them. They may also differ greatly in the way they judge the consequences of particular actions. Comparison of the different consequences can be a highly complex business.

There are considerable variations in hazard risk perception between individuals and social groups. Researchers in a number of academic disciplines (particularly psychologists, anthropologists, economists, geographers and sociologists) have tried to explain these variations in different ways. Several of the chapters in this book discuss awareness and behavioural models and factors that appear to be predictors of attitudes and actions. In general, though, there is little consensus between the various academic disciplines, and often there are variations and disputes within individual disciplines. In the introduction to their collection of key papers on theories of risk understanding, Löfstedt and Frewer (1998) point out that "Most participants in the risk debate have fundamentally different types of values and priorities which shape their definitions and judgements of risk and acceptability" (Löfstedt and Frewer 1998, pp. 4).

No single theory or model has sufficient power to explain how individuals or groups of people will perceive or respond to specific hazards and risks in all situations. Studies of risk perception – and there are now very many of these – often seem to produce inconsistent findings reflecting the extreme diversity of the environmental, socio-economic and cultural contexts in which they are carried out. However, comprehensive reviews are beginning to help us to identify some common features and themes. For example, Solberg, Rossetto and Joffe (2010) have reviewed the extensive international literature on earthquake risk perception and behavioural adjustment, highlighting the importance of cultural influences among others, whilst in this volume Lindell reviews studies of adjustments to a range of natural hazards in North America – where there has been considerable research into the subject, especially by psychologists and sociologists.

Psychologists have played a leading role in research into risk perception over several decades. As a result, much of the research in this field has tended to focus on individual perceptions of risk and the psychological factors that affect perception and lead to biases in estimation. In this volume, Johnston et al. provide a useful discussion of some of the more common factors affecting such biases, particularly relating to earthquake risk, and Karanci explains key psychological models for preparedness behaviour. Interestingly, in higher-income countries public fears of hazards seem to focus on those arising from human actions (e.g., nuclear power stations, genetically modified crops, mobile phone masts), whereas in the case of natural hazards specialists are often concerned that the public seem to underestimate the risks. There is evidence of such underestimation in this volume regarding earthquake risk perception in Japan (Yamori) and New Zealand (Johnston et al.), although the picture is different in Istanbul (Karanci).

Other researchers have placed more emphasis on the social and cultural construction of risk (Löfstedt and Frewer 1998) and as a result it is now widely acknowledged that people's understanding of risk and response to it "are determined not only by scientific information or direct physical experiences, but also by the interaction of psychological, social, cultural, institutional and political processes" (Johnston et al., this volume). In this book, Lindell observes that the social context for the adoption of hazard adjustment involves many different stakeholders and attendant power relationships. Harries' chapter explores the role of "social representations" (of the home, society and nature) in protecting people's ontological security against hazards.[1] Yamori also reminds us that societies are not monolithic in their views of risk and that many types of perception, old and new, scientific and lay, can co-exist and even interact. This simultaneous coexistence of multiple, even inconsistent, representations of recognition and knowledge known as "cognitive polyphasia" is clearly present in Japan, as Yamori shows, and presumably applies to many other places too.

Risk perception studies usually collect views at a given moment in time and rarely concern themselves with how perception might change over time and with changing circumstances. Yamori and Lindell both identify a need for more longitudinal studies.

[1] This discussion of ideas of 'home', which focuses on social representations of the home in Western societies, suggests an important line of disaster research which does not feature much in the literature on risk perception and risk management. Comparisons with non-Western societies would also be instructive.

Yamori outlines the historical development of four different social representations of earthquake risk in Japan (each of which originated at different times but all of which continue to co-exist in modern Japanese society). Johnston et al. report on a series of annual surveys of the impact of public education about natural hazards in Wellington, New Zealand.

Joffe and O'Connor call for "a departure from conventional psychological methods of experimentation and survey research" because these cannot identify the "latent emotional and socio-cultural drivers of responses to risk". By working with predefined variables, the conventional methods leave little space for new, unanticipated findings. This problem is compounded when there is a "cultural gap" between the researcher and the subjects of the research, since concepts cannot be assumed to be cross-culturally equivalent. Some attempts have been made to develop more comprehensive, flexible theories of risk perception that draw on a range of existing academic ideas. An example is the "risk compensation" or "risk thermostat" model developed by Adams (1995), which is derived partly from the anthropological cultural theory of risk: this emphasises social factors but retains elements of the individual choice models used by psychologists.

Joffe and O'Connor observe another significant limitation of conventional risk perception surveys: they "tap only consciously available cognitions", not the underlying "values, images and emotions" that may also shape people's ideas about risk. One could go further and say that, by investigating attitudes but not actions, such surveys may be inclined to overestimate the importance of words and statements or take them out of the coping context. Statements indicating fatalism – a theme which recurs several times in this book – may illustrate the limitations of surveys. In the case of earthquakes, it does appear that some kind of fatalism is common and there are indications that it can be tackled by focusing public education on the attributes of buildings rather than the power of earthquakes, thereby giving a sense that something can be done (Johnston et al.; Karanci; Joffe, Rossetto, Solberg and O'Connor in press). However, in the case of other, less extreme and more frequent events such as floods or landslides, apparently fatalistic statements may be misleading: merely conventional sayings when in fact there may be active coping strategies in place. On the basis of studies in Turkey, Karanci suggests that apparent fatalism "may be an example of emotion-focused coping", serving to reduce anxiety; Adams, considering religious responses to the 2010 Haiti earthquake, wonders if various forms of faith might be "the basis of a form of resilience" for people "condemned to fatalism by their poverty".

Some research methods can capture such strategies. For example, Jabeen and Johnson, whose research into coping with climate variability in the Dhaka slums, described in this volume, is based on qualitative interviewing together with notes, photographs and sketches, argue that amongst their sample population, fatalism "does not appear to be prevalent…insofar as people do take actions to prepare for the event and to lessen its impact on the household". Elsewhere in Bangladesh, anthropological research over an extended period of time revealed that people living on the chars (silt islands) in the Jamuna River deployed a range of methods for coping with seasonal flooding when previous conventional interview surveys had indicated they were inclined to leave everything in God's hands (Schmuck 2000). Similarly, anthropological work on risk perception in a village in the Indian Himalayas revealed that villagers' stated belief that a landslide was caused by divine wrath was not to be taken at face value since it was linked to a far more complex and subtle social construction of risk and adaptation (Pilgrim 1999).

Qualitative and quantitative evidence produce different insights into risk perception and behaviour, as Joffe and O'Connor point out. Can psychological response to risk be captured fully by way of quantitative methods, they ask, or are "important facets of the texture of thought" lost when translated into numbers? This theme still awaits full exploration but Joffe and O'Connor's discussion of the UCL EPICentre Human Loss Project shows the effectiveness of one innovative qualitative research strategy in capturing individuals' "naturalistically-occurring thought patterns" and thereby revealing the socio-cultural drivers of their responses to risk. Harries' chapter, which uses quantitative and qualitative data, further demonstrates the value of qualitative research, particularly discourse analysis, by studying the language and arguments used by UK householders to justify and explain their reasons for taking or not taking flood mitigation measures. Other work on flood risk perception in the UK has also combined quantitative surveys and qualitative interviewing effectively (e.g., Burningham et al. 2008) and this is a promising research approach.

Much of the research into risk perception has taken place in high-income countries and much of the contemporary debate on risk prediction and the so-called "risk society" is directed towards distinctly Western ways of living and thinking. How will improved risk awareness lead to change in other parts of the world where underlying socio-economic pressures compel millions of poor and vulnerable people to accept the long-term risk of natural disasters because they have to concentrate on other daily threats to their well-being and livelihoods? In academic and policy discourse, understanding of human vulnerability to disasters is increasingly based on recognition of the powerful socio-economic and institutional-political forces that shape people's lives and force them to adopt ways of living that are risky (Wisner et al. 2004).

Such groups are usually well aware of these multiple risks that they face and this guides their risk management or "coping" strategies (e.g., Twigg and Bhatt 1998; Stephens et al. 1995). Any study that considers the hazard risk perception and behaviour of individuals and groups in isolation from the broader vulnerability context in which they live can only provide a partial explanation of their views and actions. Work on hazard adaptation and coping strategies in such contexts shows that they have to be flexible above all else: poor and vulnerable people constantly have to deal with uncertainties, weigh up different options and choose actions from those alternatives that are open to them at a given time (Burton et al. 1978; Clarke Guarnizo 1992).[2]

[2] Adams' idea of the risk thermostat, in which people balance rewards and perceived danger, is not applied specifically to such groups and locations but it does encapsulate the flexibility and dynamism of coping strategies as well as pointing out the multiple interactions between each person's strategy and the complexity this creates in trying to manage risk.

Jabeen and Johnson's chapter on Dhaka, Bangladesh, is the only one in this book looking at low-income countries and the poorest and most vulnerable in them. It considers their strategies for coping with repeated, climate-related risks. Unsurprisingly, issues emerge here that are not prominent in the other chapters. In particular, Jabeen and Johnson's research sheds light "on how informal urbanisation, poverty and power relations impact on people's perceptions of hazard and the actions they take as a result". Given the proportion of the world's population that lives in the so-called "global South", in poverty and in cities where urban growth is rapid and urban development often unplanned, this chapter is of particular importance. Although a great deal of valuable research on risk awareness and preventive action has been done in higher-income and Western countries, far too little has been done in cities like Dhaka and countries like Bangladesh.

Only in Jabeen and Johnson's chapter do poverty and livelihood security play a dominant role in shaping risk perception and management. In such conditions, "people do not distinguish between hazards and other life stresses, but instead prepare for a range of possible negative events".[3] The livelihoods of Dhaka's poor, who depend on work that is labour-intensive and low-paid, often in the informal sector and without a state-supported welfare safety net, are exceptionally vulnerable to seasonal flooding. So too are their homes: in addition to being in locations that are highly exposed to floods, the householders are often squatters who lack security of tenure and face the ever-present threat of eviction. On the other hand, they have developed a range of flood coping strategies: these include higher storage facilities, increasing the height of furniture, building houses on cement plinths and making barriers to block water ingress through doorways. The range and sophistication of these active coping methods contrasts with the more passive attitudes demonstrated in some of the other chapters in this book from higher-income countries and groups – and is typical of the coping strategies used against seasonal floods by urban slum dwellers in low-income countries (e.g., Stephens et al. 1995).

10.3 Risk Communication and Popular Action

The findings and issues discussed above have significant implications for public education about risks. As Penning-Rowsell and Handmer (1990, pp. 3) point out, "Identifying new risks, deciding what is acceptable and minimising the impact of 'unacceptable' risks, requires the communication of risk information between those responsible for risk management, the risk bearers and the wider community." To be effective, such communication needs to take account of the different ways in which ordinary people perceive and respond to risks. Official British guidance for emergency planners states firmly that, "Communication which sets out to change or influence

[3] This is perhaps closer to that pre-modern orientation to peril or danger in general that Joffe and O'Connor contrast with contemporary notions that link risk with "calculation, objectivity and control".

beliefs without recognising the rational basis of those beliefs, or tries to divert attention away from people's real concerns, will almost certainly fail. A 'we know best' attitude is often a formula for disaster" (UK Resilience, undated, pp. 14).

The purpose and tone of official risk communication are derived from ideas about how risk is perceived. Where public authorities believe there is misperception or cognitive error on the part of the general public regarding risks (that it is believed scientific experts can define and calculate), the aim of public education about risk will be to correct the public's misunderstanding or fill gaps in their knowledge. This "deficit model" (Joffe and O'Connor, this volume) of perception appears to be the ideological principle behind most public education campaigns about risk. It ignores the many different rationalizations and decision making processes, psychological and social, that shape people's opinions and actions. It also seems to overlook the important fact, emphasized by Adams, that risk taking offers rewards and that the magnitude of the reward influences the propensity to take risks. Löfstedt and Frewer (1998) argue, therefore, that good risk communication should be "not a top-down form of communication from expert to the lay public, but rather a constructive dialogue between all those involved in a particular debate about risk." (Löfstedt and Frewer 1998, pp. 9). This dialogical approach, encompassing both upwards and downwards channels of communication between technical specialists and the public, is known to be effective in many areas of social welfare and development (Melkote 1991). There is also a third, informal, communications channel that should be recognized: where information is shared and discussed within families and communities (Burke 1999). This plays an important part in public response to early warnings, for instance (Nigg 1987). Social media and networking tools have added a vast new dimension to it, which disaster researchers are starting to investigate. Yamori refers to the Person Finder system made publically available by Google Japan for verifying the status of missing persons; he even goes so far as to state that, "Along with decreased reliance upon science, and fuelled by increased Internet-based communication, the amount of vernacular science knowledge concerning earthquakes and radioactive pollution has reached unprecedented levels, and blurred the boundaries between pure scientific knowledge and everyday knowledge."

Disaster managers in the twenty-first century will have to work with communities that are increasingly able to choose and question the information they receive. They will also have to move away from the traditional supply-driven model of communications, where experts at the top or centre issue information outwards and downwards to target groups. Instead, they will need to adopt a more demand-led approach that sees communities at risk as consumers of information from different sources, exercising a right to choose what information to use, where to obtain it and how to share it and apply it, in relation to their own circumstances and understandings of risk (Twigg 2003).

In the context of natural hazards, there is still relatively little understanding of the impact of public education on risk-related behaviour (other than early warnings). Yet Karanci's observation that "simple education campaigns that focus on the risks alone are not likely to significantly change behaviour" seems to be largely borne out by the literature and in this volume. There have been relatively few impact assessments and their findings vary widely: some show enhanced risk-reducing behaviour; some no effect; some even a decrease in risk perception and encouragement of risk taking. Compare for example Palm's (1981) finding that the introduction of earthquake information disclosure had minimal or no effect on Californian house buyers with Mileti and Darlington's (1995) research showing that public information initiatives were having a positive impact on knowledge and behaviour in the San Francisco Bay area. The four annual surveys in Wellington discussed in Johnston et al.'s chapter show relatively small changes, both up and down, in the proportions of people reaching certain levels of earthquake preparedness, even though awareness of earthquake risk is very high. Joffe et al. cite a research project on earthquake risk perception and response in cities in the USA, Japan, and Turkey (Joffe et al. in press) which also found that although there were high levels of awareness in each city, participants still did little to prepare for earthquakes or mitigate their risk, with emotional responses playing an important role.[4] Reviewing studies in Istanbul, Karanci concludes that low levels of preparedness and mitigation do not seem to be solely due to a perceived lack of importance of earthquakes, but rather to a much wider range of factors: "limited financial resources and knowledge on how to engage in mitigation, externalized responsibility, low trust in municipality and construction firms, denial and optimistic bias and lack of organizational support from local governments". It would seem that each communication effort and its results are highly specific to their particular location, hazards, culture, society and institutions.

Traditional civil defence and emergency management thinking has often been criticized for its preference for "command-and-control" by informed experts over co-ordination of other groups and organizations who may have differing priorities and perspectives (e.g., Dynes 2000). Emergency professionals are particularly uncomfortable with "emergent groups" – spontaneous, informal, voluntary and self-organised efforts by communities, neighbourhoods and other citizens to come to the assistance of those hit by disasters – because of their independence of thought and action, even though sociological research has shown that emergent group activity is a common and important feature in major disasters (Drabek and McEntire 2003; Stallings and Quarantelli 1985). Yamori comments on their actions in the 1995 Kobe earthquake, where there was an upsurge of volunteerism that later translated itself into the creation of disaster NGOs. Harries quotes from an interview with a flood-affected resident in the town of Reading, UK, to show the "nascent sense of belonging" that the shared experience of a hazard event can create: "Everybody was there; we were all involved in the same process of deciding whether the middle pathway was going to be closed – whether the postman would deliver or if we had to go and pick up our post. As if we were becoming a kind of a little community which was surviving an act…you know, an act of God". Johnston et al., using an example from a district in Wellington, suggest that volunteer-based community education programmes may have great potential as a means of encouraging uptake of risk reduction measures.

[4] It may be that the very high impacts and long return periods of earthquakes make them something of a special case and that we should be cautious about applying lessons from research in this field too readily to hazard risk perception in general.

Karanci discusses the relationship between improved knowledge and understanding at an individual level and people's involvement in community-based disaster reduction – the former does not necessarily lead to the latter – and calls for better analysis of the factors that facilitate joining community groups.

10.4 Responsibility, Authority and Trust

Today's decision makers have to accept that their decisions will no longer readily be taken on trust and therefore that they need to work harder to secure a sufficient degree of public confidence. Public attitudes towards those who issue risk information, and hence their confidence in the senders and the information, seem to be influenced by the public's perception of two main factors: competence and caring (Paling 2003). The more the message-senders are perceived as incompetent or self-interested, the less their messages will be trusted. A good example is the distrust shown by some Istanbul residents towards engineers and formal sector contractors, whom they suspected of being self seeking: the residents "had turned to self-construction as a means of reducing their involvement in what they perceived as a corrupt formal housing market filled with profiteering contractors, engineers and municipal building departments" (Green 2008, pp. 364). However, the picture is complex, with many varieties in the levels of trust towards different sources exhibited by different social groups. For example, there is some evidence from the USA of variations between ethnic groups in the levels of credibility or confidence they attach to different sources of disaster warning information (Perry and Lindell 1991).

Trust between citizens and public authorities is vital to efforts to promote community resilience – and hence, vital to the success or effectiveness of disaster risk management. Recent research has highlighted the importance of trust and attributions of responsibility as important factors in seismic hazard adjustment in many parts of the world (Solberg et al. 2010). Yet at present, it is by no means certain that such trust is widespread. The Turkish study conducted by Joffe and O'Connor revealed widespread distrust of state and construction authorities who were seen as negligent and corrupt. Some of the North American studies reviewed by Lindell also raise issues of trustworthiness and protection responsibility. On the basis of studies of three cities in the UK, Harries argues that "when there has been a flood, victims tend to idealise the qualities of the in-group that includes themselves and fellow victims and attribute negative qualities, including blame, to members of out-groups such as government agencies and local authorities" (and he shows how this can be overcome through effective and well thought-out community engagement strategies by disaster risk management agencies). Yamori comments on research showing that Japanese citizens are more inclined than their US counterparts to trust disaster management to government, but also shows how public confidence in scientific predictions of earthquakes and tsunami dropped suddenly after the 2011 Great East Japan Earthquake, particularly as a result of the nuclear accident at the Fukushima Daiichi plant.

A prime function of the state – if not the prime function – is to provide security for its citizens. Yet it can be argued that community resilience programmes and associated public education campaigns, by encouraging citizens to take protective actions, also sometimes run the risk of blurring the lines of responsibility for public safety. Johnston et al. refer to research showing that the dissemination of information on mitigation and preparedness options can even lead to the opposite of what is intended: a reduction in levels of household or personal preparedness and a transfer of responsibility to civic authorities. Similarly, Adams cautions, in the context of child protection systems in the UK, that "the bureaucratization of the protection of children shifts responsibility" from individuals to a more remote, impersonal system. Karanci cites a study in a neighbourhood in Istanbul which showed that one reason for local people's unwillingness to join a community-based organization working on earthquake mitigation and preparedness was that they expected state authorities to take responsibility for disaster protection measures. Harries suggests that "As long as blame is attributed to outsiders, so too is responsibility for remedying the situation". Such "externalized responsibility" (Karanci) seems to be widespread, arguably more so in countries where citizens have relatively high levels of expectation from the resources and capacities of the state – i.e., high and middle income countries – but there ought to be much more research on this in places where government authorities are perceived neither as competent nor as caring.[5] Jabeen and Johnson show that in slum settlements in Dhaka the level of household investment in actions to prepare against hazards and rebuild after disasters is closely linked to relative levels of trust – in this case, a higher level of trust in state landowners as opposed to private landowners, where the risk of sudden eviction is concerned. On the other hand, it is equally clear that the slum dwellers expect to receive no more than very limited resources and support from the government.

In the UK, the government's community resilience agenda, mentioned at the start of this chapter, is also caught up in these questions of trust and responsibility. The rhetoric of enabling can imply some shifting of responsibility, as in this fine-sounding but rather imprecise statement in the UK's community resilience framework: "the Government role is to support, empower and facilitate; ownership should always be retained by communities who have chosen to get involved in this work" (UK Cabinet Office 2011, pp. 14). Moreover, community resilience is explicitly part of the coalition government's much broader "Big Society" commitment. This ambiguous set of ideas has the ostensible vision of "a society where individuals and communities have more power and responsibility, and use it [sic] to create better neighbourhoods and local services" (Department for Communities and Local Government 2012). However, its opponents perceive it to be a smokescreen for government to abandon its duties of care for the welfare and security of its citizens – in the words of the Archbishop of Canterbury, Dr Rowan Williams, "designed to conceal a deeply damaging withdrawal of the state from its responsibilities to the

[5] The issue of trust is not confined to government: another study in Istanbul, cited by Karanci, reveals that trust in companies involved in carrying out and supervising earthquake retrofitting was a very important concern of householders.

most vulnerable" (BBC 2012).[6] Amidst this prevalent uncertainty and distrust, it would not be surprising if the new national strategy for community resilience were seen not as a way to "reduce the barriers which prevent people from being able to help themselves and to become more resilient to shocks", as the government claims (UK Cabinet Office 2011, pp. 3), but instead to be yet another instance of government passing the buck. Should this be the case, public education to heighten perceptions of risk is likely to have little impact.

References

Adams J (1995) Risk. UCL Press, London

BBC (2012) Rowan Williams dismisses big society idea as 'waffle'. http://www.bbc.co.uk/news/uk-18570664. Accessed 14 Aug 2012

Burke A (1999) Communications and development: a practical guide. Department for International Development, London

Burningham K, Fielding J, Thrush D (2008) It'll never happen to me: understanding public awareness of local flood risk. Disasters 32(2):216–238

Burton I, Kates RW, White GF (1978/1992) The environment as hazard. Oxford University Press, New York

Clarke Guarnizo C (1992) Living with hazards: communities' adjustment mechanisms in developing countries. In: Kreimer A, Munashinghe M (eds) Environmental management and urban vulnerability. World Bank, Washington, DC, pp 93–106

Department for Communities and Local Government (2012) Communities and neighbourhoods: building the big society. http://www.communities.gov.uk/communities/bigsociety/. Accessed 20 Aug 2012

Drabek TE, McEntire DA (2003) Emergent phenomena and the sociology of disaster: lessons, trends and opportunities from the research literature. Disaster Prev Manag 12(3):97–112

Dynes RR (2000) Governmental systems for disaster management. Disaster Research Center, University of Delaware, Newark (Preliminary Paper #300) http://dspace.udel.edu:8080/dspace/bitstream/handle/19716/672/PP300.pdf?sequence=1. Accessed 20 Aug 2012

Green RA (2008) Unauthorised development and seismic hazard vulnerability: a study of squatters and engineers in Istanbul, Turkey. Disasters 32(3):358–376

Joffe H, Rossetto T, Solberg C, O'Connor C (in press) Social representations of earthquakes: a study of people living in three highly seismic areas. Earthquake Spectra

Löfstedt RE, Frewer L (eds) (1998) The Earthscan reader in risk and modern society. Earthscan, London

Maskrey A (1989) Disaster mitigation: a community based approach. Oxfam, Oxford

Melkote SR (1991) Communication for development in the third world: theory and practice. Sage, New Delhi

Mileti DS, Darlington JD (1995) Societal response to revised earthquake probabilities in the San Francisco Bay Area. Int J Mass Emerg Disasters 13(2):119–145

Nigg JM (1987) Communication and behavior: organizational and individual response to warnings. In: Dynes RR, De Marchi B, Pelanda C (eds) Sociology of disasters contribution of sociology to disaster research. Franco Angeli, Milan, pp 103–117

[6] cf Harries (this volume) who sees the discourse of individual responsibility that is characteristic of neoliberal governments as an alternative means to control populations. Therefore, individual refusal to change habitual behaviour and accept responsibility for managing flood risk can be seen as a form of resistance against the state and a defence of social identity.

Paling J (2003) Strategies to help patients understand risks. Br Med J 327(7417):745–748
Palm RI (1981) Public response to earthquake hazard information. Ann Assoc Am Geogr 71(3):389–399
Penning-Rowsell E, Handmer J (1990) The changing context of risk communication. In: Handmer J, Penning-Rowsell E (eds) Hazards and the communication of risk. Gower, Aldershot, pp 3–15
Perry RW, Lindell MK (1991) The effects of ethnicity on evacuation decision-making. Int J Mass Emerg Disasters 9(1):47–68
Pilgrim NK (1999) Landslides, risk and decision-making in Kinnaur District: bridging the gap between science and public opinion. Disasters 23(1):45–65
Schmuck H (2000) "An act of Allah": religious explanations for floods in Bangladesh as survival strategy. Int J Mass Emerg Disasters 18(1):85–95
Solberg C, Rossetto T, Joffe H (2010) The social psychology of seismic hazard adjustment: re-evaluating the international literature. Nat Hazards Earth Syst Sci 10(8):1663–1677
Stallings RA, Quarantelli EL (1985) Emergent citizen groups and emergency management. Public Adm Rev 45:93–100
Stephens C, Patnaik R, Lewin S (1995) 'This is my beautiful home'. Risk perceptions towards flooding and environment in low-income urban communities: a case study in Indore, India. London School of Hygiene and Tropical Medicine, London. Unpublished report
Twigg J (2003) The human factor in early warnings: risk perception and appropriate communications. In: Zschau J, Kueppers AN (eds) Early warning systems for natural disaster reduction. Springer, Heidelberg, pp 19–26
Twigg J (2004) Disaster risk reduction: mitigation and preparedness in development and emergency programming. Overseas Development Institute, London
Twigg J, Bhatt MR (eds) (1998) Understanding vulnerability: South Asian perspectives. Intermediate Technology Publications/Duryog Nivaran, London
UK Cabinet Office (2011) Strategic National Framework on Community Resilience. Cabinet Office. London. http://www.cabinetoffice.gov.uk/sites/default/files/resources/Strategic-National-Framework-on-Community-Resilience_0.pdf. Accessed 20 Aug 2012
UK Resilience (undated) Communicating risk. Cabinet office, London. http://www.cabinetoffice.gov.uk/sites/default/files/resources/communicating-risk-guidance.pdf. Accessed 22 Aug 2012
UNISDR (2005) Hyogo framework for action 2005–2015: building the resilience of nations and communities to disasters. UN International Strategy for Disaster Reduction, Geneva. http://www.unisdr.org/2005/wcdr/intergover/official-doc/L-docs/Hyogo-framework-for-action-english.pdf. Accessed 14 Aug 2012
Wisner B, Blaikie P, Cannon T, Davis I (2004) At risk: natural hazards, people's vulnerability and disasters. Routledge, London

Index

H. Joffe et al. (eds.), *Cities at Risk: Living with Perils in the 21st Century,* Advances in Natural and Technological Hazards Research 33,
DOI 10.1007/978-94-007-6184-1, © Springer Science+Business Media Dordrecht 2013

MIX
Papier aus verantwortungsvollen Quellen
Paper from responsible sources
FSC® C105338

If you have any concerns about our products,
you can contact us on
ProductSafety@springernature.com

In case Publisher is established outside the EU,
the EU authorized representative is:
Springer Nature Customer Service Center GmbH
Europaplatz 3, 69115 Heidelberg, Germany

Printed by Libri Plureos GmbH
in Hamburg, Germany